Inside CERN's Large Hadron Collider

From the Proton to the Higgs Boson

Inside CERN's Large Hadron Collider

FROM THE PROTON TO THE HIGGS BOSON

Mario Campanelli

University College London, UK

W World Scientific

NEW JERSEY • LONDON • SINGAPORE • BEIJING • SHANGHAI • HONG KONG • TAIPEI • CHENNAI

Published by

World Scientific Publishing Co. Pte. Ltd.

5 Toh Tuck Link, Singapore 596224

USA office: 27 Warren Street, Suite 401-402, Hackensack, NJ 07601

UK office: 57 Shelton Street, Covent Garden, London WC2H 9HE

Library of Congress Cataloging-in-Publication Data
Campanelli, M. (Mario), author.
  Inside CERN's Large Hadron Collider : from the proton to the Higgs boson / Mario Campanelli (University College London, UK).
    pages cm
  Includes index.
  ISBN 978-9814656641 (hardcover : alk. paper) -- ISBN 978-9814656658 (softcover : alk. paper)
  1. Large Hadron Collider (France and Switzerland). 2. Colliders (Nuclear physics). 3. Particles (Nuclear physics). 4. Higgs bosons. 5. European Organization for Nuclear Research--History.
I. Title.
  QC787.P73C36 2015
  539.7'36--dc23
                                          2015023671

British Library Cataloguing-in-Publication Data
A catalogue record for this book is available from the British Library.

In-house Editor: Ng Kah Fee

Typeset by Stallion Press
Email: enquiries@stallionpress.com

Contents

Chapter 1

Introduction

Always men struggled to make sense of the surrounding world. Are the stars in the sky the results of old fights between gods and heroes? Are the mountains, the rocks, the rivers, coming from primordial creatures? Is mankind at the centre of the world? Could the complexity of the world surrounding us be expressed by complex interactions of simple elements?

Many ancient cultures have the concept of four elements (air, water, earth, fire), whose combination could describe the whole universe as well as the human body, through the corresponding four humours. Isn't the human body, after all, a small universe in itself?

Alchemists tried over centuries to convert various states of matter between them, with the ultimate goal of transforming everything into gold. More materials were discovered, and the complexity seemed overwhelming. However the idea of understanding our universe with simple elements was always there, and was the basic guidance of centuries of investigation into the structure of matter, of hopes, of failed attempts.

The real breakthrough arrived just a few centuries ago, a very short time compared not only to the existence of the human species, but also to the history of rational thought.

The use of mathematics as a tool to understand the world, combined with the experimental scientific method, changed our world, and the conception of our universe.

The abolition of the principle of authority, the systematic doubt, the taste for the new, on one side replaced the Earth from the centre of the universe and man from its special status, putting it at its place inside the animal world; on another, the path was opened for the formidable adventure of understanding the nature of matter and the structure of the universe in a quantitative way, a way that allows today to make very precise predictions of quantities that a few decades ago were not even known.

The adventure of fundamental physics made us realise that the law of nature at microscopic level are completely different from those we are used to in daily life; made us rethink of fundamental concepts like space, time and causality. The recent discovery of the Higgs boson, and the measurement of its mass, mean that our universe may actually lie in a state of meta-stability.

All this seems very abstract, but is also very concrete: quantum mechanics is at the basics of modern electronics through the semiconductor transistor; lasers are used every day in optical disks and supermarket scanners; general relativity is even used in the calculations performed to make GPS devices accurate at street level.

Despite, or probably thanks to, its enormous success, modern fundamental physics is challenged with more questions than ever. The current theory, that is so much in agreement with thousands of experimental results to be commonly called the "Standard Model", still does not explain gravity, the matter–antimatter asymmetry of the universe, and the rotational speed of galaxies, just to name a few. Now that cosmological and astrophysical observations have also reached higher and higher precisions, the connections between measurements made at very big and very small scales become stronger and stronger.

More precise measurements are needed, new landscapes have to be explored. Big scientific discoveries happen when new measurement challenges the accepted view, and forces us to rethink what we already know.

In this context, thousands of scientists from all over the world joined their efforts to build the largest scientific instrument ever built: the Large Hadron Collider, a 27 km-long proton and ion accelerator, and its four gigantic particle detectors. This machine is the obvious follow-up of decades of accelerator-based particle physics, and the tipping point of centuries of studies of the structure of the matter and of fundamental interactions.

This book will describe the main steps that have led us to the current understanding of particle physics, before digging into the details of the accelerators, the detectors, the men and women behind them, and the major results obtained so far.

Chapter 2

The Development of Particle Physics

Men of different cultures and traditions developed models of the surrounding world that helped them to make sense of what was happening, and to predict the consequences of their actions. Often these models were inaccurate, connected to unproven myths. Most of them are lost, with perhaps some echoes surviving through the oral tradition.

With the development of agriculture, the need for measuring land, and for making some astronomical predictions to regulate the cycles of work in the fields, has led to the development of mathematics, and abstract thought in general. Mankind started to ask questions about the nature of things, and our destiny, in a more rational way, even if for many centuries myth and rational thought were still strongly connected.

Very few fragments still exist of pre-Socratic Greek philosophers, traditionally considered the first heroic attempts in the western world of trying to understand nature using rationality, for the pure sake of it.

Reading these old fragments can be very illuminating and moving: some of the questions were the same as those of today, others would now be considered naive; in all cases, the effort to understand was mixed with the marvel of the complexity of nature.

Many approaches were followed; some would be called in modern language more "theory-inspired", others more based on the observation.

A concept that remained in the collective imagination, also because it is shared by many other Eastern cultures, is that of the four elements. By observing pockets of air trapped in a bucket partially filled with water, Empedocles proved that air is indeed a separate element, and while previous thinkers considered a single element as the root of everything else, he formulated the theory that the universe was actually a combination of four "roots" — air, water, earth and fire. Clearly, none of them would today be considered as a fundamental element (pure water is a molecule of just two elements; air is a mixture, dominated by three gases; earth has a very complex and variable composition; and fire is actually a process, not a material!), but the idea was powerful: perhaps the complexity of all matter is just apparent; perhaps, matter is just a complex mixture of simple elements!

The idea was developed by others, especially Demokritos, who coined the term "atom", that means un-splittable, so basically a building block of all matter. In his view, many atoms with different properties, falling over the Earth, would make the complexity of the matter.

They would be abstract objects, not representing concrete materials like air or water. Of course no experimental proof could be made at these times of the real existence of such "atoms". In parallel, the different approach of observing the natural world to derive similarities has led to the precise classification of animals and plants culminating in the late works of Aristotle and Theophrastus. Between these "purely speculative" and "purely empiric" approaches, some amazingly modern synthesis emerged, like for instance the almost correct prediction of the Earth's radius made by Eratosthenes, or the alleged prediction of a solar eclipse by Thales. These two examples show that Greek philosophers/scientists were not only able to postulate or describe, but to produce a synthesis, and make falsifiable predictions, a fundamental aspect of modern science.

For many centuries the exploration of nature was more episodic than systematic, and often based too much on the principle of authority than on the direct observation of nature.

For centuries, alchemy produced large bodies of experimental evidence, without being able to produce a synthesis. Using the scientific method, modern chemistry managed to produce in just a few decades results that would change our perception of the world forever.

In 1779 Lavoisier discovered that water, one of Empedocle's elements, is actually made of two elements, called hydrogen and oxygen. Many other elements were discovered, while some materials, like gold or iron, did not seem to be composed of different constituents.

Following the reductionist approach, the search for fundamental building blocks of nature began on a scientific basis. Elements were classified according to their properties, and in the early 19th century Dalton discovered that in making compounds, they reacted according to proportions given by integer numbers. He interpreted this fact as a confirmation that each element contained a fundamental brick, an "atom", that contained all chemical properties of the element itself, and that it was the combination of the atoms that produced new compound materials. Shortly afterwards, Avogadro proposed that two equal volumes of different gases, at equal temperature and pressure, should contain the same number of atoms, and observing that sometimes in naturally-occurring gases like nitrogen or oxygen this number was doubled, he concluded that in nature they are organised in diatomic molecules.

Shortly it was found out that all existing matter, be it in solid, liquid or gas state, is made of combinations of just 92 elements, that is, 92 types of atoms. These elements were classified according to increasing atomic weight and their chemical properties in Mendeleev's periodic table. The atomic theory had reached its triumph, when already the regularities observed in the disposition of the elements in this table made scientists think that perhaps atoms were not that un-splittable, but that smaller constituents could explain these similarities between atoms of very different size.

Section 2.1: First elementary particles

In addition to the regularities observed between the various atoms in the periodic table, the nature of the chemical bounds was difficult to explain without assuming substructure of atoms. The answer came from a field apparently quite far from chemistry: the study of electricity.

Already by the end of the 19th century, Maxwell's equations had unified electricity and magnetism: in other words, phenomena as different as, for example, electric current flowing into a wire and the attraction between a magnet and a piece of iron, could be explained by an elegant set of four equations. Both electric and magnetic forces were proportional to a quantity called electric charge, which could apparently take any possible value. But perhaps all electric charges were multiples of a still unknown very small fundamental charge? Perhaps electricity was carried by a very small particle?

Two fundamental experiments clarified these issues, one by J.J. Thomson in Cambridge in 1895, the other by Millikan in Chicago in 1909.

Thomson, who was the first Nobel Prize winner for Physics, observed that heating up a metallic filament produced electrically charged particles, which could be accelerated (therefore deflected on a screen) by electric and magnetic fields. These particles that are found in atoms had all the same ratio between electric charge and mass, so presumably were all identical. They did carry electricity, and were called for this reason "electrons". Atoms were not fundamental building blocks after all.

The era of particle physics had started.

Millikan measured that the electric charges of very small droplets were small multiples of the same number, which he correctly identified as the charge of the electron. Combining the information of the charge with that of the charge over mass ratio from Thomson's experiment, the electron mass was found to be a few thousandths of the mass of the atom containing them.

Electrons were therefore responsible for only a very small fraction of the mass of an atom, but for all its negative charge. Additional matter had to be present in atoms, responsible for almost all of their mass; since most of the atoms are neutral, at least part of this additional matter had to carry positive electric charge to compensate the negative charge of the electrons. Since measurements of the inter-atomic distances in crystal lattice indicated orders of magnitude of the size of an atom of about 10^{-10} metres, Thomson himself postulated that atoms were made of a positively charged substance, with electrons immersed in it like the raisins in a plum pudding. Matter had, according to this model, an approximately uniform distribution of positive and negative charge. Other models were proposed, notably the one where all positive charge and most of the mass of atoms are concentrated in a central nucleus, with the electrons orbiting it like planets in a solar system. The advantage of this model, proposed by Ernst Rutherford, is that the external electrons of nearby atoms could interact among themselves, creating chemical bonds. Rutherford even proposed an experiment, realised by his collaborators Geiger (later known for the development of a radiation counter) and Marsden, to discriminate between the two models. At the time, with the exploration of the atomic structure at its beginnings, it was not obvious how to explore dimensions of the scale of 10^{-10} m, and this approach became the conceptual basis of many other scattering experiments performed in the decades to follow. The concept was simple: to know how something is built, bombard it!

The experiment consisted in bombarding the atoms of a thin foil of gold with α particles, and measuring the scattering angle. Rutherford himself had discovered that these particles were none other than helium nuclei, with twice the opposite of the electron charge. The idea is that if matter has a uniform distribution of positive and negative charges, the electric field felt by the α particles will be on average close to zero, and these particles cannot be deviated by too large angles.

If on the other hand the positive charge is concentrated in a central nucleus, a particle getting close to this nucleus would feel a very strong field and undergo large deviations. Making a classical parallel, if one shoots a bullet through even a very large plum pudding, the bullet will exit approximately with the same direction. If instead the pudding is replaced by a small lead ingot with the same mass but much larger density (well, some puddings can be quite dense as well, especially around Christmas!), when the bullet hits the ingot it could be deviated at very large angles. And this is what Geiger and Marsden

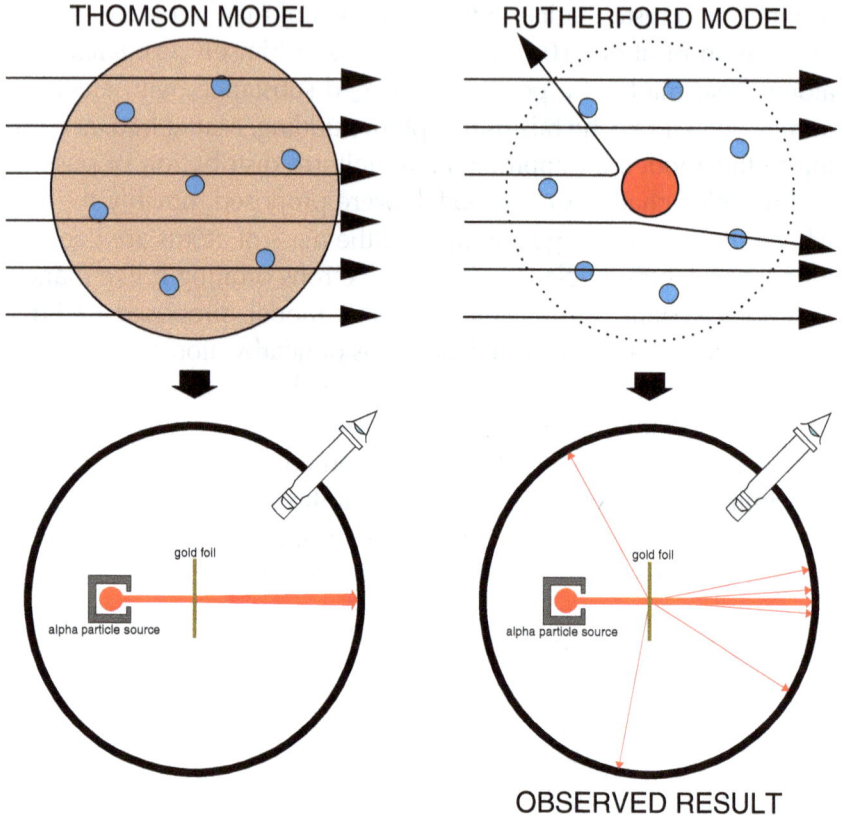

Figure 1: Schematic description of Rutherford–Geiger–Marsden experiment: in a model with distributed positive charge (left), projectiles experience very small deviations, while if the positive charge is concentrated in a small central region (right), deviations can be much larger. (Image credit: Kurzon, Wikipedia)

observed, to much delight of Rutherford, who is now considered the father of the modern atom, and buried in Westminster next to Isaac Newton.

Thomson's "plum-pudding" model was therefore discarded, and the conceptual picture of atoms as a small solar system, as we see in many images, was born. It is however worth noticing that the image of electrons like little spheres moving in circular or quasi-circular orbits around a central nucleus is also not correct.

A charged particle moving on a circular orbit, or in any way undergoing a change with respect to a straight, constant-speed trajectory, will emit electromagnetic radiation, losing part of its energy. This would cause the electrons to fall on the nucleus in a very short time. The electrons inside atoms, on the other hand, have constant energy, and do not fall on the nuclei. Clearly the laws of physics are different at the microscopic and macroscopic scales, and in fact the stability of the electron orbits is one of the paradoxes that have led to the development of quantum mechanics. A more accurate way to build an image of the electrons orbiting a nucleus is to think of strangely-shaped clouds of charge, or rather clouds of probability of finding a charge, around the nucleus.

What about the nucleus then? Its radius is a hundredth of a thousandth that of the atom, but it contains more than 99.9% of its mass. It must contain positive particles, but what else?

If we order the atoms according to their atomic weight, like in Mendeleev's table, we can assume that nearby elements differ by just one electron, but the mass is not proportional to the electron number. Moreover, there are some atoms with slightly different masses, but identical chemical properties, indicating the same electron number and structure. The obvious solution is that nuclei are made of two kinds of particles: positive protons, in equal number to the electrons (at least for neutral atoms), but with about 2000 times their mass; and neutral neutrons, with similar mass to that of the protons.

Materials with the same number of protons and electrons, but different number of neutrons, will have almost identical chemical properties but different mass.

All the elements of the periodic table, so the large majority of the matter surrounding us, are combinations of just three particles: protons, neutrons and electrons; reductionism has never been stronger.

So, if everything can be explained by just three particles, whose properties are by now very well known, why are we still doing particle physics? The key is in the "large majority": not everything can be explained by these three particles.

In fact, by performing experiments with cosmic rays and the first particle accelerators, the physicists of the early 20th century discovered a real zoo of particles: pions, muons, kaons, lambda... particles with very different and interesting properties, but something in common: a very short lifetime, meaning that after a very short time (of the order of a millionth of a second, or less) they would decay, transformed into something else. Protons, electrons and neutrons (inside nuclei) are instead stable, at least from what we know; in any case their lifetime is longer than the lifetime of the universe, and this is why they are so abundant.

But why are some particles stable, while others have such short lifetimes? In the disintegration chain, heavy particles decay into lighter ones, which in turn decay again and again, until the lightest particle of its category is produced, and it will be stable. Protons, neutrons and electrons are the lightest particles of their category, and this is why they are stable, and so abundant.

To better understand the relation between energy and mass of particles, and what it really means to discuss "particles" or "waves", we have to briefly discuss the main principles and implications of the two great conceptual revolutions of fundamental physics in the first half of the 20th century: relativity and quantum mechanics.

Section 2.2: Special relativity — a new concept of space and time

As already briefly mentioned, a great success of classical physics was the unification of electricity and magnetism: current flowing in a conductor produces a magnetic field, and similarly the change of a magnetic field flux through a circuit can induce electric currents. From the equations describing these phenomena it can be inferred that electromagnetic radiation, i.e. combinations of rapidly varying electric and magnetic fields, can travel in space at very high speed. The use of this radiation to transmit information has changed the world: radio, television, long-distance communications are all con-sequences of this big fundamental physics discovery.

An interesting issue is that the equations also provide the speed at which this radiation propagates in vacuum: the speed of light, c, which is a fundamental constant that is the same for any electromagnetic radiation, corresponding to 299,792,458 m/s.

We know from our daily experience that speed depends on the reference frame: if we are on a car running at 100 km/h, overtaking another one running at 80 km/h, the speed of the first car as seen by the second will be 20 km/h. All that corresponds to our daily experience, and is called classical or Galilean relativity: every speed depends on the reference frame with respect to which it is measured. However, the electromagnetic theory just gives the speed of light in vacuum, without reference to any reference frame. Attempts to define a privileged frame (ether) for the propagation of electromagnetic radiation was excluded by the experiments of Michelson and Morley as early as 1887.

Moreover, if the speed of light was depending on the reference frame, in moving along the same direction as light one would see it slowing down, even stopping. In this case one would be able to "ride" a light beam, and that has never been observed.

The solution came from an eccentric employee of Bern's patent office, whose name was destined to enter popular culture as that of the archetypal scientist: Albert Einstein.

He was born in Germany, but moved to Switzerland to attend university. Partly because of his unconventional attitude, he was not considered good enough to get a scholarship to do his PhD at ETH Zurich, and had to pay tuition fees by himself. At the end of the doctorate, no perspectives for academic jobs opened, and he took a job in the patent office. Fortunately for the course of modern physics, the job was not very demanding, and he had time to continue working on some open problems of physics of his time.

In 1905, he published three articles, each of them worthy of the Nobel Prize (which he will only receive in 1921, for the photoelectric effect, which we will discuss in the next section about quantum mechanics). The article on relativity addresses exactly the issue of the speed of the electromagnetic radiation, and will change forever our conception of space and time.

Instead of postulating a special reference frame for the propagation of the electromagnetic radiation, relativity assumes that its speed is the same in every reference frame; that all electromagnetic radiation travels with the same speed, while no object with mass can ever reach this speed. That assumption is perfectly coherent with Maxwell's equations, and solves the issue of not being able to "ride a beam of light", but leads to very non-intuitive consequences. The same term "relativity" has been largely misinterpreted in the vulgarisation, leading to statements like "everything is relative" that have little physical meaning. Einstein himself later said that he should have called this theory "invariance" rather than "relativity".

If the speed of light is a constant in every reference frame, the simple sum of speeds between reference frames cannot be true any more: it is only a low-speed approximation.

Let us take again the case of the car, moving at 80 km/h, overtaken by another car moving at 100 km/h, which will be seen by the first one to have a relative speed of 20 km/h.

Now, if the first car is replaced by a spaceship, moving at the incredible speed of 80% of the speed of light, and the second one by a beam of light, moving at the speed of light c, one would naively expect, in analogy with the case of the cars, the crew of the spaceship to observe the beam moving at a relative speed of 20% of the speed of light.

Einstein's relativity is telling us that this is not the case: observed by the spaceship, the beam of light will always move at the same speed, c, as if it was observed by a stationary position. Actually, the concept itself of stationary reference frame loses its meaning: all reference frames are equally relevant, and electromagnetic radiation moves at the same speed in all of them. This leads to other non-intuitive consequences.

Let us suppose in the spaceship there are two mirrors, facing each other, and parallel to the direction of motion, such that a bored astronaut could send a light signal on one mirror, and have it go back and forth between the two, perpendicular to the direction of motion. The time between successive reflections of the light signal between the mirrors is given by their distance divided by the speed of light, and this time interval could be used as a basis to build a clock. An observer, for which the spaceship moves at 80% of the speed of light, could see the same scene. For him the distance travelled by the light signal is much longer: it will not only depend on the distance between the mirrors, but also on the fact that the mirrors themselves are actually moving, so it will be longer. But if the speed of light is the same as the one measured inside the spaceship, the time interval between two reflections, as observed from outside, will be longer. The observer will see that time in the spaceship flows at a slower pace to him, while the people in the spaceship will observe nothing unusual. In relativity we lose the concept of a universal time, of a universal "cosmic clock" beating the hours for everybody in the same manner. The flow of time does depend on the relative speeds of the reference frames, as if each reference frame had its own clock. As there is no universal time, the scale of distances also depends on the reference frame.

If time flows at a different pace in the two reference frames, but the speed of light is the same, a distance defined as the space travelled by light in a unit time (different between the two system) will also be different between the two. All this is quite in contradiction with our everyday experience: after all, distances and time seem to be the same in all reference frames, and in classical physics there is no concept of speed-dependent space or time. The point is that all these relativistic effects are only relevant at speed differences close to the speed of light; this almost never happens in everyday life, and this is why relativity was discovered so late in the history of physics. A proof of the validity of time dilation comes from the fact that we observe cosmic muons at sea level. These particles, similar to electrons but heavier, decay into electrons and neutrinos after an average time of two millionth of a second. Even travelling very close to the speed of light, muons could only travel about 600 metres before decaying. Cosmic muons are produced in showers initiated by collisions of primary protons about 10 km above sea level. Even accounting for late production, muons should not be able to reach sea level in the quantities we observe. The answer is precisely in the time dilation: the muon lifetime is two millionth of a second in the particle's reference frame, but since the muon is moving at speeds very close to that of light, his lifetime in our system can be much longer. Or, thinking in terms of space, the atmosphere looks much "thinner" if seen in the reference frame of a muon. However you should not be mistaken and try to move very fast (however unrealistic and super-expensive it would be to travel at speeds close to that of light) to live longer: the muon, like any system, has always the same lifetime in its own reference frame, it is only with respect to another external frame that the pace of time slows down.

Another very important consequence of special relativity is the relation between mass and energy. In classical mechanics, the kinetic energy of a particle is given by half its mass multiplied by its speed squared. There is no maximal speed, and particles can reach any energy. In relativity, a massive particle can never reach, let alone exceed, the speed of light, while a massless particle will always travel at the speed of light. The energy can still grow without any upper bound in both cases. The relation between energy, velocity and mass becomes more complex, and is such that as particles get closer to the speed of light, even enormous increases in energy result in very small fractional variations of the particle velocity. Particle accelerators can reach speeds very close to that of light, so relativity is tested in laboratories like CERN on a daily basis. As an example, protons enter the Large Hadron Collider (LHC) at an energy of 450 GeV and are accelerated to 7000 GeV, an increase by almost a factor of 20. However, the speed increases from 99.9997% to 99.999999% of the speed of light: it gets closer to the limit speed, but it will never reach it.

Another very interesting consequence of the relativistic relation between mass and energy is that a particle at rest, so with zero speed, still has some energy due to its mass: it is the famous relation $E = mc^2$, meaning that whenever a heavy particle decays into some lighter ones, some kinetic energy is liberated (this is the working principle of a nuclear power plant), or that massive particles can be produced from energy. This is exactly the reason why we build large particle accelerators: standard particles are given an enormous kinetic energy and made collide; the kinetic energy is transformed into mass, and new, heavy particles are created. The relation between energy and rest mass also explains why in our universe only protons, neutrons and electrons are stable: our universe is the result of a cooling-down that lasted 13 billions of years, and the average inter-galactic temperature is about −270 °C; low temperature means low energy, not enough to keep in thermal equilibrium anything other than the lightest particles of their category.

Section 2.3: Quantum mechanics — a world of particles, waves and uncertainties

In our everyday world some phenomena are specific of individual objects, like the collisions between billiard balls, while others are collective, like the movement of waves in the sea made of enormous amounts of individual droplets.

Quantum mechanics, one of the most important conceptual revolutions of modern science, is telling us that this separation is just illusory, and that if we look close enough, or small enough, there is a single nature, exhibiting features of both waves and particles.

Light certainly behaves like waves: it can be reflected by a mirror, and a narrow slit exposed to it will become a source of cylindrical waves, like the circular ones produced when large waves from the sea hit a narrow opening of a harbour.

Throwing stones in a lake will produce circular wavelets that will interfere with each other, producing intricate patterns of maxima and minima. In the same way, monochromatic lights interfere and produce interference maxima and minima. However, light can be seen as behaving very much like particles: Einstein did not get his Nobel Prize for the relativity theory, but for explaining the photoelectric effect. It was observed that electrons can be extracted from a material by shining some laser light on it; however, no matter the intensity of the light, the phenomenon could be observed only if the frequency of this light was above a given value. This is easy to explain if we consider light to be made of particles, called photons, each with energy proportional to the frequency (with the same proportionality constant as used by Planck to explain the black body radiation some years before!); extracting an electron is not a collective effect, but the action of each individual photon: no matter the total number of photons (i.e. the intensity of the light), if none of them had enough energy to extract the electron, no current could be produced.

Similarly, electrons have all the characteristics that we would attribute to particles: they have a mass, a charge, and they interact individually. However, if we send electrons through a slit that is narrow enough, the same diffraction and interference phenomena observed for light and mechanical waves are observed.

We have already seen that the electrons in atoms do not radiate, so cannot be considered really moving in an orbit like little planets of a solar system. These electrons can only occupy very specific energy levels that correspond, in a semi-classical view, to the position of a wave with circumference equal to a multiple of its wavelength therefore constructively interfering with itself. It is interesting to notice that, while ancient astronomers were convinced that the orbits of the planets had to follow some very regular patterns (the harmonies of the spheres) when there is in fact no regularity, such integer relations have been found by modern physicists by looking inside the atoms!

So, electrons, photons, and similarly all other particles have both properties of waves and particles: the electron around the atomic nucleus is not a small planet, but can be seen as a wave, whose local intensity is connected to the probability of finding the charge in a given position, or with a given momentum, around the nucleus. The only quantity we can predict is the probability, and even if a single measurement, for instance of the electron position, gives a well-defined answer, a series of measurements will produce a probability distribution that can be compared to the theoretical predictions, intrinsically probabilistic. This is why, for instance, we need to run our accelerators for several years and collect billions of collisions, in order to make precision measurements and strict theory comparisons. The position and momentum of a particle are however strictly correlated, to the point that it is not possible to measure both with infinite precision. The famous uncertainty principle states that the product of the uncertainties on position and momentum cannot be smaller than the Planck constant divided by 4π. This is a very small number at our macroscopic scale, but relevant at atomic scale; while we do not have to worry about the uncertainty principle to measure the position and speed of a tennis ball, because anyway the precision of our instruments will never be better than the intrinsic limitation, in measuring the position and momentum of an electron of an atom the uncertainty principle plays a much larger role. So, if you get a speeding ticket, do not tell the policemen that the speed of your car was uncertain: yes, it was, but you were probably above the speed limits by much more than the tiny uncertainties due to quantum mechanics!

The interconnections between energy and momentum, together with the probabilistic nature of quantum mechanics, depict a microscopic world very different from the one we are used to in everyday life. But if we look a bit deeper, things are not so different. For instance, the uncertainty principle already exists in wave mechanics. If we play a note, say the A from a diapason, the nominal 440 Hz frequency of that note will be defined with infinite precision only if the diapason plays for an infinite time. If, as it happens in the real world, the note is only played for a limited duration, it will be a superimposition of frequencies, centred around 440 Hz, but with a spread inversely proportional to the duration of the note. Since the frequency is connected to the momentum, and for a travelling wave the duration is connected to the position, the product of the uncertainties of these two quantities can never go to zero. While already present in classical physics, the uncertainty principle does not occupy there the central role it has in quantum mechanics; it is the fact that particles also have a wave-like nature that makes it so much more relevant for the interpretation of the microscopic world.

For macroscopic systems, composed of billions of billions of particles, not only the uncertainty principle plays no role given the size of the relevant measurements, but also all other quantum effects (apart from very specific, coherent systems) average out, leading to the apparently deterministic, and familiar, laws of classical physics.

So if in everyday life quantum effects are basically absent, why bother to understand them?

Well, first of all, because of curiosity: we want, we have to understand the world around us; it is there, as Edmund Hillary answered to those asking him why he wanted to climb Mount Everest. A more subtle answer is that even the behaviour of microscopic objects could teach us a lot about our macroscopic world, and lead to profound technological innovations. Quantum mechanics is the basis of the greatest technological innovations of our time, like lasers and modern electronics. Having a semiconductor chip hosting billions of transistors in a square centimetre is an incredible accomplishment, deeply rooted in the understanding of modern physics: the inventors of the modern transistor, Bardeen, Brattain and Shockley, were quantum physicists, and received the Nobel Prize in Physics in 1956. Without this discovery, all modern computers, cell phones, digital cameras would be impossible (or maybe enormous and with limited performance, like the first valve computers from the fifties).

The development of quantum mechanics was a long adventure, and its philosophical implications are still under debate. A big step forward was the development of the quantum theory of the electron by Paul Dirac, which extended quantum mechanics to particles with spin, and described by Einstein's special relativity. The merging of the revolutionary ideas of relativity and quantum mechanics opened a completely unexpected development, a door to another world, or perhaps it would be better to call it an anti-world.

Section 2.4: Antimatter

Various attempts to use relativistic dynamics in the equations of quantum mechanics all led to solutions that could be interpreted as particles with negative energies, or going backward in time. The most complete formulation that also incorporates spin was completed by Dirac in 1928, and the negative-energy solutions were still there. Dirac correctly interpreted those as particles identical to the ordinary ones in all their properties, apart from some which would be opposite, like the electric charge. These particles are called antiparticles, or particles of antimatter. When a particle meets its antimatter correspondent, the two would annihilate, with all their mass converted into energy, typically in the form of light. So Dirac asked various astronomers if the observation of the sky showed any sign of these annihilations, of photons with energy identical to that of the mass of the known particles. The answer was negative: no signs of antimatter in the nearby universe. Dirac started to believe that perhaps the antimatter solution was just a mathematical trick, with no physical reality. Fortunately there are also experimental physicists, and Carl Anderson discovered in 1926 antielectrons in cosmic rays, with a beautiful apparatus consisting of a cloud chamber in a magnetic field, and a basic trigger system connected to a camera. The direction of curvature of the track in the magnetic field gave a measurement of the charge, but there was still an ambiguity between downward-going negative particles and upward-going positive ones. This was brilliantly solved by placing a slab of lead in the middle of the chamber; the particle would be slowed down by the lead, and present a smaller radius of curvature after it; that allowed to determine the direction of the particles, and therefore establish the discovery of antimatter. This experiment is a beautiful example of a relatively simple and small apparatus able to make a major fundamental physics discovery, something we miss a lot in today's big-science world.

After this experiment, antimatter partners for all known particles were found, as well as helium antinuclei. Today we can even produce antihydrogen atoms in a laboratory. Still, the search for antimatter in the universe, also performed with very sophisticated tools like the AMS detector orbiting on the International Space Station, only found small quantities of antimatter in cosmic rays and a belt of anti-electrons trapped in the Earth's magnetic field, but no evidence of large quantities of antimatter, like anti-stars or anti-galaxies. This is surprising, since we assume that the universe originated from an explosion where energy produced matter and antimatter in equal amounts. A theory from A. Sacharov (better known later as a soviet dissident) from 1967 showed that if some conditions are met, among which differences in the laws of physics between matter and antimatter, a symmetric universe may turn into an asymmetric one, and therefore the almost complete absence of antimatter could be the result of all of it having been annihilated, and our universe would be made of just the small percentage of matter that survived this enormous annihilation.

An asymmetry between the behaviour of matter and antimatter has been found in weak interactions in 1964, and since then several experiments have been performed (among which LHCb at the LHC) to study this phenomenon; the general consensus now is that the amount of asymmetry found is way too small to explain the current universe, so the search for additional sources of this asymmetry continues.

Section 2.5: Particle zoo

Cosmic ray experiments not only discovered antimatter, but a whole series of new particles. One of them had a mass about 200 times larger than that of the electron, so it was initially called "meson" as its mass was intermediate between that of the electron and of the proton. Initially, this particle was thought to be responsible for strong nuclear interactions, as predicted by Yukawa, but there was a little problem: this particle did not feel the strong nuclear interaction at all! It was later understood that the particle believed to carry the strong interaction (we know now that things are more complicated than this) was the pion, and the first meson, now called the muon, is a heavy electron. The scientific community, with the voice of I. Rabi, asked the question: who ordered the muon? In other terms, why nature decided to replicate itself in families, with particles that are copies of the known ones, but just heavier?

But more surprises had to come: studies with cosmic rays and with the first accelerators discovered a huge number of additional particles that were given exotic names like kaons, lambda, sigma… all of them with a very short lifetime, therefore very rare in our universe.

With the discovery of all these new states, people realised that the number of "elementary" particles was becoming too large for all of them to be really elementary, especially since these particles had several regularities, like the atoms of the periodic table. So maybe all these particles were not elementary after all, but made of something smaller, like the atoms of the periodic table are actually made of smaller protons, neutrons and electrons?

This question was approached by two sides: experimental and theoretical. Experimentally, scattering experiments (conceptually similar to Rutherford's experiment that unveiled the structure of the atom) where high-energy electrons were sent through fixed targets demonstrated that the proton could be broken into smaller constituents, since after the collision other particles (mainly pions) were emitted. Does it mean that protons are made of pions? Or perhaps that both protons and pions are made of something even smaller?

Theoretical physicists, among which Murray Gell-Mann, observed the symmetries between quantum numbers of the various particles, and classified them into diagrams. These symmetries could not just come from chance, and the hypothesis was made that these particles were not really elementary, but were made of smaller constituents, which Gell-Mann called "quarks", from a quote of James Joyce's Finnegans Wake.

It was not difficult to understand that indeed the quark model could explain the breaking up of protons in scattering experiments. Protons, like pions and all the many particles that feel the strong nuclear force are indeed bound states of (at least at first order) either three quarks, or a pair of a quark and an antiquark. The strong nuclear interaction forces these quarks to always stay together, so no free quarks can be observed, but only these combinations. When an electron was hitting a proton, the quarks in the proton would be "scrambled", and immediately after would recombine into other stable particles, among which pions. As far as we know, on the other hand, electrons are indeed elementary particles: no experiment has ever managed to "break" an electron into smaller constituents.

Section 2.6: Neutrinos

Beta decays are a well-known process where neutrons in some unstable nuclei get transformed into electrons and protons. Free neutrons undergo beta decays after an average lifetime of 15 minutes; if the neutrons had not combined with protons to form stable nuclei in the first minutes of the life of the universe, they would all be decayed by now.

A puzzling property of beta decays is that, even when the neutron was at rest, the momentum of the outgoing proton did not balance that of the outgoing electron. Initially at rest, the system acquired a momentum, which is of course forbidden by the law of momentum conservation. Also, the total energy plus mass of the electron–proton final state is smaller than the mass of the initial neutron. This decay apparently violates both energy and momentum conservation, two fundamental laws of classical physics!

The solution (and, as it often happens, the beginning of many more questions) came in 1930 from Wolfgang Pauli. Since he could not attend a conference in Tubingen where these issues were discussed — the annual ball of his university was held in Zurich on the same day — he sent a letter, famously addressed to "dear radioactive ladies and gentlemen". In this letter he claimed that the apparent non-conservation of energy and momentum could be solved if the assumption was made that in the beta decay an additional particle was emitted together with the proton and the electron. This particle needed to be neutral, invisible, have almost no interaction with ordinary matter, and have very small mass. Pauli himself once said: "I did a terrible thing: I postulated the existence of a particle that cannot be observed!" He called it the neutron, but then confusion arose with the neutrons present inside the atomic nuclei. Enrico Fermi, in his mathematical description of beta decays, found a nicer name for this new particle: he called it neutrino, since it had to be lighter than what we now call neutron; and the suffix -ino, in Italian, is a diminutive (like violino, violin, is a small viola, the Italian name for alto).

Pauli was wrong in thinking that this particle was undetectable, but it still took about 25 years before neutrinos could actually be discovered by Cowans and Reines using a detector placed next to the Savannah River nuclear reactor. Billions of neutrinos were produced in the nuclear fission reactions, and only a handful of them, as predicted by Fermi's calculation, could indeed interact with the detector and produce a detectable signal. The first thing they did was sending a cable to Pauli, and probably for once the old professor was happy to be proven wrong.

Even if we do not see them, and they have little influence on our everyday life due to their very small interaction probability, neutrinos are produced in an enormous variety of processes. Low-energy neutrinos fill the universe since they were produced in enormous quantities during the early phase of its evolution; closer to us, the nuclear fusion reactions of the Sun emit neutrinos that we can today detect and measure; cosmic rays produce hadronic showers that also contain neutrinos as their final states; supernova explosions produce enormous quantities of neutrinos, that can reach our planet and be detected. But neutrinos are also produced in more familiar settings: radioactive decays, in small doses, are much more common than what we think, and an average man, since our body contains traces of radioactive elements, produces a few thousand neutrinos per second; a banana, which contains potassium, emits about ten neutrinos per second! However, even if many experiments have been performed to measure neutrinos from astrophysical sources, nobody has ever managed (and probably even tried!) to measure neutrinos emitted by a banana!

Successive experiments demonstrated that indeed neutrinos come in three species. Those emitted in beta decays, in nuclear reactions and in the Sun produce electrons when they interact with matter. Most of the neutrinos from cosmic rays, and the artificial neutrino beams we can create in accelerators, produce muons when they interact. Some other neutrinos, the last to be discovered, even produce tau leptons when they interact. So, as we have three charged leptons: electrons, muons and taus, with very similar characteristics but with different masses, there are also three neutrinos families, each for a corresponding charged lepton.

But what about the mass? To account for the apparent non-conservation of energy and momentum in beta decays, neutrinos have to be very light, but the precision of the measurements could not rule out the existence of a very small but non-zero mass. In the fifties, Bruno Pontecorvo demonstrated that if neutrinos have even a small mass, they could oscillate, namely a neutrino of one family could spontaneously convert into one of another family. So muon neutrinos produced in an accelerator can, after few milliseconds (corresponding to hundreds of kilometres, since given their very small mass, neutrinos travel basically at the speed of light), become electron or tau neutrinos. The oscillation is a periodic phenomenon: if after a given average distance a muon neutrino becomes an electron neutrino, after twice that distance it does oscillate back into a muon neutrino, and so on.

The first indication of possible neutrino oscillations came from the Sun. Over the years, scientists (among them John Bahcall) developed a model of the Sun that predicted the number and energy of electron neutrinos that our star should produce in its various reactions. Very difficult and ingenuous experiments managed to detect these low-energy solar neutrinos, but their number was about half of what the model predicted. Of course this could indicate a problem with the model, but a much more interesting possibility emerged: that perhaps the model was right, but on average half of the electron neutrinos from the Sun oscillate into muon neutrinos during their travel to the Earth. Since solar neutrinos have a much smaller energy than the muon mass, they cannot produce a muon when they interact with matter, and the neutrinos that oscillated into muon neutrinos remain undetected. Only the other half of the neutrinos, consisting of those that after the oscillation came back to the electron state, were measured by the detector, leaving the false impression that the Sun was producing half of the neutrinos that it indeed was. Finally the SNO detector managed to measure also reactions sensitive to the presence of muon and tau neutrinos, confirming that the total neutrino flux coming from the Sun was in agreement with the model predictions, and therefore the factor of two drop was due to oscillations.

Proof of neutrino oscillations also comes from measuring in big underground laboratories the neutrinos produced by decays of cosmic ray particles, and from neutrino beams. High-energy neutrinos can be produced by the decays of pions from accelerators, and sent to detectors located hundreds of kilometres away, to study the oscillation phenomenon in a controlled way. Neutrino beams have been sent from CERN to Gran Sasso in Italy, from Fermilab (Chicago) to Soudan (Minnesota) in the USA, and from J-PARC (north of Tokio) to Kamioka (in the Japanese Alps) in Japan. Given the tiny interaction of neutrinos with matter, no tunnel is needed for the neutrino beam. The neutrinos will cross hundreds of kilometres below the Earth's crust, reach the faraway laboratory, and most of them will never interact there, but just get out of the Earth again. Basically, we are sending out neutrino beams in the cosmos, like particle lighthouses.

To summarise, the three neutrinos, and their charged correspondents, the electron, the muon and the tau, are considered to be elementary particles, belonging to the class of leptons. These particles do not feel the strong nuclear interaction, and are found isolated in nature. Another class of particles are quarks, also divided into three families: the first contains up and down quarks, the second charm and strange, the third top and bottom. Quarks do feel the strong nuclear interaction, and are forced by it to always mix between them. We will never see free quarks, but always mixtures of them. Combinations of the six quarks can produce hundreds of particles, called hadrons. Protons and neutrons are in fact combinations of up and down quarks. Each quark and lepton also has an antimatter correspondent, bringing to 24 the total number of elementary building blocks of matter.

Section 2.7: Fundamental forces

But there is more: Physics not only deals with matter, but also with forces that make this matter interact. Classical Physics had always tried to describe the largest number of natural phenomena with the smallest number of forces. One of the first successes of modern science has been the unification of terrestrial and celestial gravity, in Newton's universal gravitation. For us it is obvious that the same force responsible for the falling of the apple is the one that makes the Moon orbit around the Earth and the Earth around the Sun, but at the time it was a revolution: the laws of nature were the same on Earth and in the sky! A simple equation could explain all the motion of the solar system and more, something impossible using the assumption that orbits had to be spherical, just because the sphere was a "perfect" solid.

After Newton's universal gravitation, several other unifications followed in classical physics. Thermodynamics was unified with statistical mechanics: we know today that temperature is none other than a faster or slower movement of atoms and molecules in a gas or a solid body. Electricity was unified with magnetism: very different phenomena like the current flowing in a wire, or the attraction between a magnet and a piece of iron, can be explained by the four Maxwell's equations. Many other phenomena, like electromagnetic induction, or electrostatics, can be explained by these equations; moreover starting from them the physicists were able to understand and use electromagnetic waves, which are rapidly varying (and travelling) electric and magnetic fields.

Around 1950, all natural phenomena could be described by just four fundamental forces.

Gravity is probably the most familiar to our experience. Newton's theory has been completed by Einstein's theory of relativity that describes the attraction between two massive bodies in terms of curvature of space and time. This theory is able to predict observed deviations from the behaviour described by Newton's gravity, and is of fundamental importance to understand cosmology and the evolution of the universe. However, general relativity is a classical theory, it does not account for quantum-mechanical effects, and scientists think that an even more general theory exists, which has quantum mechanics and general relativity as subsets.

The electromagnetic force, responsible for many familiar processes like light, friction, chemical bounds etc., has been described in quantum-mechanical terms by the theory of Quantum Electro-Dynamics (QED), for which Feynman, Schwinger and Tomonaga received the 1965 Nobel Prize. This theory has been tested with probably the highest accuracy in the history of science: the electron magnetic moment has been measured, and correctly predicted by theory, with a precision of a few millionths of a millionth.

The weak nuclear force is responsible for several radioactive decays like the beta nuclear decay and the decay of the muon. It is also the only force to be different between matter and antimatter. It is by its nature a quantum-mechanical force, and described for the first time in these terms by Fermi.

The strong nuclear force is what forces the quarks to always mix between them to produce more stable hadrons, and protons and nucleons to be bound together in atomic nuclei. Initially thought to be carried by pions, now the strong force is described with high accuracy by the theory of Quantum Chromo-Dynamics (QCD), for which Gross, Politzer and Wilczek were awarded the 2004 Nobel Prize. This theory is similar in its formulation to QED, but has more technical complications that limit the precision of some predictions.

A vast theoretical effort has culminated in the sixties into the unification of the electromagnetic and the weak interaction, in what is today called the electroweak theory. Glashow, Weinberg and Salam received in 1979 the Nobel Prize in Physics for this unified theory, even though, as it often happens, many other scientists did contribute. This theory received an enormous amount of experimental confirmations, and it is fair to say that today we have not yet found any convincing measurement that is in contradiction with it.

To better understand how this unified theory works, we need to know that in quantum mechanics, a force is the result of a particle exchange. For instance, the electromagnetic interaction is carried by photons: each time two electrons repel each other, they exchange a photon — we can think that the first electron emits the photon and moves away, the second electron receives this photon and is also moved away by absorbing it. Of course this view is too simplistic, since it cannot for instance describe attractive forces, but the intuitive picture is quite correct.

Particles that transport forces are called "bosons" (from the name of the Indian physicist Bose), while those composing ordinary matter, like leptons and quarks, are called "fermions" (from Fermi). They differ in their spin, which is a property connected to the angular momentum that can be imagined as having a particle turning around an axis. Bosons have integer spin, like 0, 1 or 2, while fermions have semi-integer spin (actually, all fundamental fermions have spin $\frac{1}{2}$).

We already said that the electromagnetic force is carried by photons, quanta of light. In a similar way, the strong nuclear force is carried by gluons (so called since they "glue" together quarks in more stable hadrons), and the weak nuclear force is carried by two more bosons: the Z (with charge zero) and the W (that can have positive or negative charge). We can think of the electroweak theory as a unification of the particles carrying the two interactions; in fact, the photon and the Z behave like two separate particles only under some circumstances, but the electroweak interaction is carried by a mix of the two. They interact with quarks and leptons, and are organised under very precise symmetry groups.

Why was the photon discovered much earlier than the Z boson, and why is electromagnetism much more common in our daily experience than the weak interaction? The reason is that the photon has zero mass, while the Z "weights" almost like 100 protons, so producing a photon is energetically almost for free, while the Z requires much more energy. The mix between photon and Z that is exchanged in electroweak interactions depends on the energy (actually, on the momentum transferred in the interaction), so for most of the common interactions happening at low energy, the messenger of the electroweak interaction can be approximated by a photon, and the interaction itself approximated by a purely electromagnetic one. At low energies, purely weak interactions have a small probability of happening, hence the name.

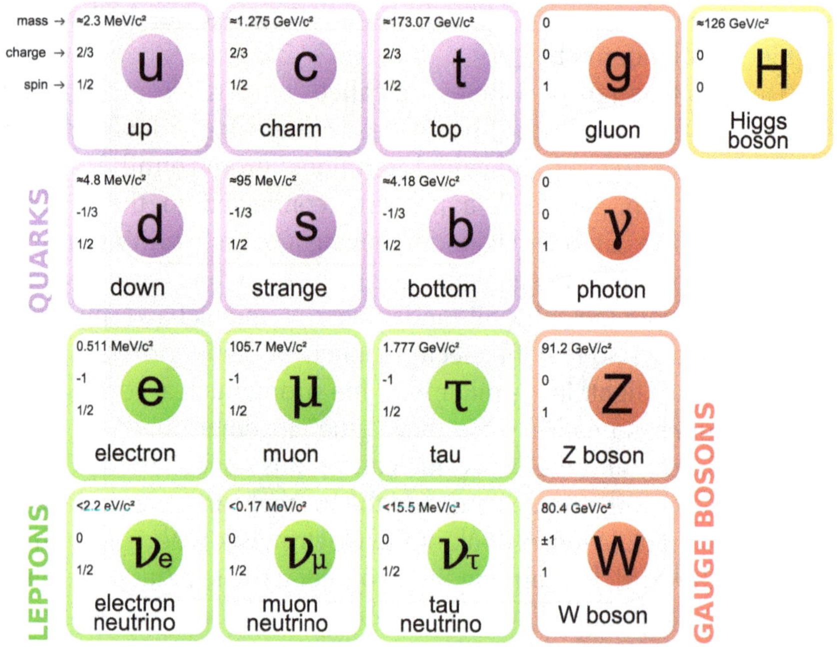

Figure 2: Summary table of the elementary particles as we know them now. Particles are divided into fermions and bosons, with fermions composing ordinary matter and bosons carrying forces, or the interaction responsible for mass. (Image credit: MissMJ, Wikipedia)

Section 2.8: Higgs boson

The electroweak theory assumes that W and Z bosons interact with leptons and quarks of all families in the same way. It is based on symmetry principles, and all particles (including the W and Z bosons) are intrinsically massless. This is sharply in contrast with the observations, which show that almost all particles have a mass, even very different between them. It is not possible to accommodate a direct mass term in the equations of the electroweak theory, it would not make any mathematical sense. The solution is much more elegant: the mass emerges as the result of an interaction with an external field. As we have in the universe the electromagnetic field and the gravitational field, there is also another field, called the Higgs field. Each particle interacts with the Higgs field, in other words it exchanges constantly energy with this field. This interaction breaks the symmetry of the theory, and this is what gives mass to otherwise massless particles. Each kind of particle has a different interaction strength with the Higgs field, and this explains why masses of various types of particles are all so different.

To get a pictorial image, we could say that since the mass is just a form of energy, what we call mass of otherwise massless particles is none other than the energy that they constantly exchange with the Higgs field. Moreover, as in quantum theory fields are carried by particles, the Higgs field is also the result of the exchange of a particle, the Higgs boson. The Higgs field can interact with itself, so the Higgs boson can also acquire a mass by the same mechanism.

The mathematical formulation of the theory is more complex than just particles getting their mass by interacting with a field. The symmetry breaking is like a little sphere on top of a sombrero. When the ball is at the highest point, it is fully symmetric under rotations of the sombrero, but it is unstable. Realistically, a small perturbation will change this situation: the ball will fall and find a stable equilibrium position somewhere at the bottom of the hat. But the new position will not be symmetric any more: the ball will have a special position. Similarly, it is thought that the breaking of the symmetry of the electroweak theory happened during the first phases of the life of the universe, where a non-symmetric solution allowed the mechanism to give mass to otherwise massless particles.

Even though the Higgs field is everywhere, and so the Higgs bosons, producing real Higgs bosons, with enough energy to create all its mass, is very difficult, and it took decades to actually prove the existence of this particle and measure its properties.

The Large Electron–Positron Collider (LEP), operational at CERN between 1989 and 2000, searched for the Higgs boson produced in association with a Z boson. It could therefore place a limit close to its maximal energy (about 209 GeV) minus the mass of the Z boson (91 GeV); the final lower limit was about 114 GeV: if the Higgs boson exists, its mass had to be larger than 114 GeV, otherwise it would have been discovered at LEP.

The Tevatron accelerator at Fermilab, operational for over 25 years between 1983 and 2011, was colliding protons and antiprotons with total energy of about 2 TeV, so much more than LEP, but with a much higher background noise. Finally, Tevatron could exclude a Higgs mass window between 160 and 170 GeV; an excess was observed in the region around 125 GeV, but the significance of this signal was not enough to claim a discovery.

The search for the Higgs boson is one of the goals of the LHC, even if by no means the only one. During its first years of operation, the LHC excluded the existence of the Higgs in a mass window between 140 and 450 GeV, so combining this result with the one from LEP, either the Higgs was very heavy (a case strongly disfavoured by theory), or had to be found in the most difficult region, the one between 114 and 140 GeV.

After many years of searching, in July 2012 the discovery of a particle with properties compatible with those of the Higgs and a mass of about 126 GeV has been announced by the ATLAS and CMS experiments at CERN.

Section 2.9: Future prospects of particle physics

The Higgs discovery does not mark the end of the open questions in particle physics, it actually opens new ones. We believe that, since it does not include gravity and does not really unify the strong interactions in a coherent framework, the electroweak unified theory, also known as the Standard Model, is only an effective theory, valid at relatively low energies, but incorporated in some larger model. In a similar way, Newton's mechanics is a theory perfectly valid at low speeds (where low in this context means with respect to the speed of light), but it is just an approximation of relativistic mechanics, which is valid even at relativistic speeds.

Many models have been proposed for what "new physics", beyond what we know already, should look like. Most of them try to go in the direction of further unifications, or overcome the limitations of the Standard Model itself. Since we do not have yet a proper description of gravity in quantum language, and due to its very weak interaction strength, it is believed that the strong nuclear force should be the first to unify with the electroweak.

Few years after the development of the electroweak theory, models were formulated suggesting a unified description of the strong and the electroweak force, using an extension of the mathematical formulations that were so successful in unifying the weak and the electromagnetic interaction. If these models were theoretically valid, they would predict the possibility of protons and neutrons decaying, with lifetimes of about 10^{32} years, much longer than the current age of the universe that is about 10^{10} years (more precisely, about 14 billion years). Even if the predicted lifetime is so large, it is incompatible with the universe as we know it now. More refined versions of these theories predict even longer lifetimes for these particles, and several experiments have been looking for proton and neutron decays, usually using very large volumes of pure water placed hundreds of metres underground, and continuously observed by thousands of electronic eyes, ready to spot decay signals of any of the nucleons in the water. So far, no indication of proton decay has ever been observed, and therefore we do not have any experimental evidence of the validity of these Grand Unified Theories (GUT).

But even if we do not aim for further unifications, the electroweak model has some strange behaviour. For instance, if new particles are present at higher masses than currently probed, the value of the

Higgs mass would be influenced by these additional particles, and would be much higher than what has been measured, unless cancellations considered unnatural occur. And we do expect new heavy particles to exist, since that would solve one of the longest-standing problems in particle physics, that of dark matter.

Already in 1932 it was observed that the outer parts of galaxies move at a higher speed than what would be predicted by the observed mass distribution. In fact, the vast majority of the mass of the galaxies seems to be concentrated in a central core, like in the solar system most of the mass is concentrated in the Sun. But in the solar system the inner planets, like Mercury, move at much higher angular speeds than the outer ones; in galaxies, the rotational speed is instead quite uniform, making people think that there is indeed in galaxies much more mass than observed. Decades of studies of this phenomenon, combined with cosmological observations and studies of gravity at large scales suggest now that the most likely explanation for this phenomenon is the existence of new particles, which are either heavy or weakly interacting or, most likely, both. Light particles interacting in the early phases of the universe with protons and neutrons would have led to a density much more uniform than we observe.

If dark matter particles are heavy, the issue of the smallness of the Higgs mass has to be addressed. It can be explained if we postulate that for each known fermion there is a corresponding, and not yet observed, boson, and likewise for each known boson there is a fermion. Note that this symmetry between bosons and fermions (also known as supersymmetry, or SUSY) is not the same as that between matter and antimatter, but yet another one, leading to the existence of a whole new group of particles. SUSY-inspired theories have been a very active field of study for many years, and still are one of the most favourite object of investigation at the LHC. Probably discovering SUSY would be even more important than having discovered the Higgs; moreover, supersymmetric theories predict the existence of at least five Higgs bosons! It has to be however pointed out that to stabilise the Higgs mass, SUSY particles should not be too dissimilar in mass to their Standard Model correspondents, so if at the end of the LHC program no indication of the existence of these particles is found, many physicists think that the theoretical appeal of this approach may be greatly reduced.

Another point of view is to assume that new particles have not been found not because they are too heavy, but because their coupling is too small, and this could also explain why, even if their mass is small, their presence has not influenced the formation of the large-scale structures of the universe. Many of these models can be studied at the LHC, but for the most interesting parameter values it would be more suitable to build other dedicated experiments, perhaps with smaller energy but larger collision intensity.

Many other approaches to new physics exist: some assume that quarks and leptons are not elementary particles, but are made of smaller constituents; some propose the existence of particles with strong coupling to leptons and quarks; others postulate new interactions or new symmetries.

The quest for symmetry is very common in particle physics: it has been a guiding principle in the development of the quark model, of the electroweak theory, of grand unifications with the strong force, just to name a few; it is at the basis of many more models. Why do physicists consider symmetry to be so important, a guide in the exploration of the new? Certainly in part aesthetic considerations play a role. In part history has shown symmetry to work as a guiding principle, as long as mechanisms to break it are in place. In part the quest for symmetry also comes from the Big Bang theory. We know that the universe few moments after the Big Bang was much hotter, so much more energetic, than today. It was not only full of very heavy particles, but also much more symmetric. At very high energies, the difference in mass between an electron and a muon, for instance, becomes negligible, and these particles behave in a very similar way. Also their abundance in the early, very energetic universe was very similar. Then the universe expanded, and like the gas in a fridge it cooled down. As the average energy went below the mass of the muon, there was not enough energy in the universe for muons to be in thermal equilibrium, and the majority of them decayed. In our current, cold universe, the difference in mass between electrons and muons is very important: the electrons are everywhere, while muons are much rarer. We believe that the use of symmetry will get us closer to the basic laws of physics, and to the primordial universe.

The discussion about the relative abundance of muons and electrons can be extended to many other particles. The primordial universe was much more interesting than today, since it was filled with heavy exotic particles, which later disintegrated during the cooling down of the universe without being replaced. Today, the only truly abundant particles are the lightest of their category: electrons, protons, neutrons and electron neutrinos. With this in mind it is easy to understand the mistake of the scientists of the early 20th century, calling them "elementary particles", convinced to have found the true building blocks of nature. Perhaps in the future some surprise will show us that what we know today is just the tip of the iceberg of a much more complex reality.

Chapter 3

A Short History of CERN

The name CERN is often associated with advanced technology and state-of-the-art science, but the history of this lab is quite long, and CERN is one of the first inter-European organisations created after the Second World War. Let us go back to those times: Europe was rebuilding itself from the debris of this terrible war where neighbouring countries have been invading and bombarding each other for years. Resources for fundamental science were scarce, and Europe, especially the countries where dictatorships were in power before the war, was plagued by a strong brain drain: many scientists, some very famous ones but also many others that constitute the fundamental scientific "humus" for the advancement of research, had left the continent because of political or racial persecutions. But fortunately a few enlightened scientists had remained (among them Bohr, Amaldi and Auger), and they understood something very important: if the various European countries wanted to avoid becoming irrelevant in the new post-war world, they had to forget the rivalries and start to build a common project. So, something incredible happened: countries that had been bombarding each other until a few years earlier began to collaborate in a very ambitious peaceful scientific endeavour.

Already in 1949 the proposal of having an inter-European scientific collaboration was presented at a cultural European conference.

The following year, UNESCO accepted a resolution proposed by the nuclear physicist I. Rabi to promote the creation of international scientific laboratories, and the Conseil Européen de la Recherche Nucléaire (European Council for Nuclear Research) was funded. The following year the location was established in Geneva, because it was at the centre of Europe; Switzerland had been neutral during the war, and it was already the location for many international organisations like the UN, the Red Cross, etc.

However, with the memories of the nuclear bombs that had destroyed entire cities at the end of the war, concerns were present about the creation of a "nuclear physics" laboratory; a referendum against it was proposed, but it was rejected by the Geneva citizens.

In the following year, officially on October 19, 1954, the laboratory was created, and it kept the name CERN, from the council.

Section 3.1: The accelerators

If the administrative progression was impressive, the scientific one was even more so. After two years, the first accelerator, the SynchroCyclotron (SC) started operations, accelerating protons up to 600 MeV. This machine was operational until 1990. But the laboratory had a much more ambitious plan: in 1959 the Proton Synchrotron (PS) started operations, and until today this amazing machine is still at the heart of CERN's accelerator complex, after having accelerated almost continuously for over 50 years protons, electrons, heavy ions and many of their antimatter correspondents.

The initial laboratory was located in a plot of land in Switzerland right at the border with France; in the 60's, the French government donates the neighbouring area to the organisation, which became the first international organisation physically located across two countries. The new land allowed the construction of the Intersecting Storage Rings (ISR) — a double accelerator where two proton beams, circulating in opposite directions, could collide and therefore have much higher energies than those of a single beam hitting a fixed target. That was the first time that the idea of counter-rotating colliding beams, already used in other laboratories for electrons and positrons, was applied to protons, like it is now in the LHC. The ISR took data between 1971 and 1984, with several achievements (among which, for instance, the first proton–antiproton collisions), but lacking big scientific discoveries.

An important discovery came in 1973, when the Gargamelle bubble chamber observed events where high-energy neutrinos interacted without producing charged leptons in the final state. Until then, a muon neutrino beam interacting in a detector had always produced final-state muons, in a process known as charge current exchange (because a neutral neutrino was transformed into a charged muon). The absence of the muon, and the analysis of the collision, showed that another neutrino was present in the final state, and therefore the process had to be mediated by a neutral current, or actually a neutral particle.

That was predicted by the unified electroweak theory, for which two particles could be exchanged in the weak interactions, typical of neutrinos: the W boson, responsible for charged currents, and the Z boson, responsible for neutral currents, not yet observed at the time. Neutral currents are common in electromagnetism, so having observed them in weak interactions was the first convincing experimental proof that the two forces could be described by a unified theory.

Unfortunately the scientist in charge of this experiment, Lagarrigue, died in 1975, so he never received the Nobel Prize that this discovery would have deserved.

While the observation of neutral currents was an indirect proof of the existence of the Z bosons, the neutrino beam did not have enough energy to actually produce this particle. For that another hadronic collider was needed, namely the transformation of the next large machine of CERN, the Super Proton Synchrotron (SPS). Inaugurated in 1976, this gigantic accelerator (7 km in length) was built at a depth of about 100 metres, largely in neighbouring France. A second surface laboratory was built entirely on French land. Just after this machine started, a proposal came to operate it in collider mode, accelerating protons in one direction and antiprotons in the opposite one. This was never done before, mainly because producing antiprotons is a complicated and inefficient process, and the few antiprotons produced have a wide range of energy and directions, hardly the ideal case of the very collimated and mono-energetic beams that can be accelerated in a collider! Only the invention of the "stochastic cooling" technique, that forced antiprotons produced in a wide range of energies and directions to become more aligned and mono-energetic, allowed to have a large collision rate.

In 1983, the two large experiments, UA1 and UA2, placed around the collision points of the collider, recorded the first collisions that could be identified as the indication of real W and Z bosons being produced. In the following year (a record time, marking the importance of the discovery), Carlo Rubbia, who together with David Cline proposed running the machine in collider mode, and Simon van der Meer, who invented the stochastic cooling, were awarded the Nobel Prize in Physics. The electroweak unification received a spectacular confirmation: the properties of these bosons were exactly those predicted by the theory.

This experiment could only produce a handful of these particles, enough to claim their discovery and study the main properties, but not enough to allow precision measurements. That was possible with the next ambitious project of CERN: the construction of the largest particle accelerator in history. LEP was a gigantic (27 km) ring where electrons and antielectrons circulated at speeds very close to that of light, and collided in four points, around which gigantic detectors (ALEPH, DELPHI, L3 and OPAL) collected the products of the collisions. During its construction in the 80's, the building site of the LEP tunnel was the largest civil engineering project in Europe; but it was worth it: during its operation, between 1989 and 2000, LEP produced millions of Z bosons and tens of thousands of W bosons, allowing the study of the electroweak theory with an unprecedented precision and the search for new physics in new energy regimes. If the search for new physics could not find positive results, the measurements of the parameters of the electroweak theory will probably remain unsurpassed for yet many years to come.

Yet, a piece was missing in the otherwise very well detailed puzzle of the electroweak model: the origin of mass. The model predicts that mass is the result of an interaction with an external Higgs field, and this field is carried by a particle, the Higgs boson. However, the LEP experiments could not find this particle, despite a lot of effort, and placed a lower limit of 114 GeV on its mass. To continue the search for this fundamental particle and to explore many other models of new physics, even higher energy was needed: a new, more powerful accelerator had to be built.

Electrons rotating in a ring lose enormous amounts of energy due to synchrotron radiation, so the high energy required to produce heavy particles can only be achieved by building a proton collider. It was decided to use the same 27 km tunnel that hosted LEP to build a new machine able to reach energies about 70 times those of LEP: the Large Hadron Collider. After almost ten years of preparation, in late 2008 the first beams started circulating in the LHC, but after a few days a problem in the interconnections forced the machine to stop. The LHC was to take the first data in 2009, and it kept increasing its energy until reaching 8 TeV in 2012, the year CERN announced to the world the discovery of the Higgs boson, after many years of searching. LHC had a two-year shutdown between 2013 and 2015, and has restarted in June 2015 at the energy of 13 TeV.

Section 3.2: Other experiments

The history of CERN is not only that of big accelerators. Smaller machines can also give important contributions to the understanding of matter and its properties, and the richness of the lab has always been the diversity of its research program.

We have seen that neutrino interactions in Gargamelle were a fundamental stepping stone in confirming the validity of the electroweak theory. Other neutrino experiments explored the structure of the proton and studied neutrino oscillations, the phenomenon in which neutrinos of one kind spontaneously transform into neutrinos of another kind. Since oscillation can occur in milliseconds, and these particles travel at speeds practically indistinguishable from that of light, oscillated neutrinos have to be detected kilometres away from production. This is why an experiment has been set up to send neutrinos to the Gran Sasso laboratory, about 100 km east of Rome, at 730 km of distance from Geneva. Fortunately, no tunnel is needed: neutrinos have such a small interaction probability with matter that those lost in the trip are a very small fraction; unfortunately, that also means the number of neutrinos interacting in the Gran Sasso laboratory is very small, about one per day, compared to the billions of neutrinos per second produced at CERN. The experiment aimed at discovering the oscillation of neutrinos from the muon to the tau family, a process with a probability of a fraction of a percent. Coupled with the very small neutrino interaction probability, the result of this experiment has been the observation of just a handful of tau neutrinos in several years of measurements; few, but sufficient to establish the existence of this rare oscillation mode.

Another important field of study at CERN are the properties of antimatter, especially bound states. The first atoms of antihydrogen (an atom made of an antiproton and an antielectron) were produced in 1995 by sending xenon gas against an antiproton beam circulating in a storage ring. After that seminal discovery, a new experimental facility was set up, where intense antiproton beams can be produced and can be trapped in vacuum in magnetic containers; there they can mix with antielectrons, producing relatively large amounts of antihydrogen atoms. The goal is to measure the energy levels of antielectrons in these atoms, and compare them to those of hydrogen; current theories predict that no differences between matter and antimatter atoms should be observed, but we hope that nature could surprise us once more.

Section 3.3: An example of technology development — the invention of the web

CERN is not only about pure science, but also technology development. In the nineties, a new idea, started as a communication tool for particle physicists, changed the world: the World Wide Web. To be clear: internet, as the infrastructure allowing the communication of computers all over the world, exists since the fifties, and was invented in the US for military purposes. But until the invention of the web, very little public information was available over the internet, and it could only be accessed using complicated commands, such that it was reserved only to some computer specialists. The great idea of T. Berners-Lee was to develop a system to allow anyone to produce public pages and post them on the internet, create links between the pages, and allow people to move between pages and share documents and figures with a click. The web changed the world, but was created as a collaborative tool between particle physicists: scientists working on CERN experiments are located in universities all over the world, and the internet was a great way to have them exchange information, all that was needed was an easy tool to post and retrieve this information. Certainly the original pioneers of CERN could not foresee how this tool developed to exchange information between physicists would have changed the world.

But the incredible story of the web can also be read as a metaphor of the interplay of pure and applied research taking place at CERN: in the last decades, the lab has been searching for the Higgs boson; in the meantime the web was invented, and finally also the Higgs boson has been found. Pure and applied researches always go together in an amazing place like CERN.

Chapter 4

The Challenges of the LHC

The largest scientific instrument in the world is located about 100 metres underground, between Geneva airport, in Switzerland, and the Jura mountain range, in France. The tunnel hosting it was built in the eighties for the LEP accelerator; LEP was then completely removed from the tunnel in the year 2000, and then the new machine was built. The idea of having a proton collider inside the LEP tunnel existed since a very long time, even before the start of LEP itself, and for some years it was thought possible to have the two machines co-existing in the same tunnel, to also perform electron–proton collision; however the excessive weight of the LHC magnets made it impossible to have them placed above LEP, as initially thought, and the cryogenic system ended up requiring much more space than initially foreseen, forcing the removal of LEP.

In the LHC a few thousand bunches of protons, each containing 100 billion particles, are turning in opposite directions at speeds close to that of light, colliding 40 million times per second at four collision points. Around these points, four giant detectors have been placed: ATLAS, CMS, LHCb and ALICE. Each time two bunches collide, an average of 30 protons collide with another 30, making a total rate of over a billion collisions per second.

The accelerator runs about 10 months per year, 24 hours a day and 7 days a week. This does not mean that there are continuous

collisions, but almost: apart from periods of maintenance and accelerator studies lasting a few days, preparation phases of 4–5 hours are
followed by 15–20 hours of collisions; then a new preparation period
follows, and new collisions, etc.

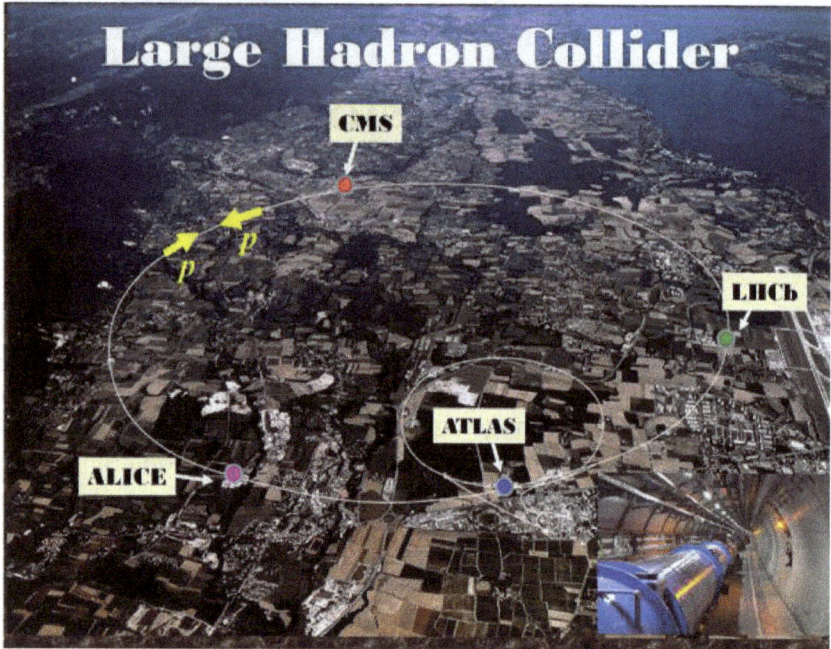

Figure 3: Aerial view of the region between Lake Geneva and the Jura Mountains,
with circles showing the position of the two main accelerators of CERN: the SPS (the
smaller ring near bottom right) and the LHC, also showing the position of the four
detectors.

Section 4.1: Superconducting magnets

To understand how an accelerator works we have to go back to the
basic laws of electromagnetism. A charged particle in an electromagnetic field feels two forces, both proportional to the charge of the
particle: the electric force, going in the same direction as the electric
field; and the magnetic force, perpendicular to both the direction of
the magnetic field and to that of the particle.

A charged particle in a uniform electric field will behave like a mass in the Earth's gravitational field: it will be accelerated, and will receive energy from the field. In an accelerator rather than uniform and constant electric fields we use electromagnetic waves to give energy through electric fields to the particles in accelerating cavities.

A particle in a uniform magnetic field, on the other hand, will change its direction. Since the force and displacement are perpendicular, the force will not perform any work on the particle, therefore it will not contribute to its energy. Magnets (at least those producing a magnetic field independent of time, like those we find in accelerators) are not giving any energy to the particles; they are used to guide their trajectory, like the rails of a train. The apparently strange characteristic of the LHC is that out of 27 km there are only about 30 metres of cavities; only one part per thousand of the length of the machine is made of a real accelerator, the vast majority of its length is reserved for magnets used to guide the particles in their orbit.

There are 1232 15-metre long dipole magnets (the size is close to the maximum allowed length of a truck allowed on European motorways), where an 8 tesla magnetic field (about 150,000 times the value of the Earth's field) bends the beam directions by a few mm every 15 metres. To obtain an 8 tesla magnetic field with an electromagnet, a coil with 50 windings in a few cm is needed, with currents in excess of 10,000 amperes. This is a big number, since typical values for a domestic installation are of the order of 20–30 amperes. Carrying over 10,000 amperes with a normal copper cable is possible, but very unpractical: very large cables are needed, possibly with a powerful water cooling system, and enormous amounts of energy are lost to warm up and then cool down the cables. Copper magnets would be enormous, and difficult to build within the necessary mechanical tolerance. Fortunately an alternative exists, and is the use of superconductivity.

Superconductivity is the property of some materials to lose completely their electrical resistance at very low temperatures (plus some additional conditions). In the LHC a current of 13,000 amperes flows in cables with a width of one cm for a thickness of just one mm, made of a special alloy of niobium and titanium. This material becomes a superconductor at temperatures just a few degrees above the absolute zero, and in fact the LHC is the largest (and one of the coldest) refrigerator in the world: all 27 km are cooled down to the temperature of 1.9 absolute degrees (−271°C), a temperature colder than that of the intergalactic space.

The cooling of the magnets is performed by flowing inside the magnets liquid nitrogen for the first phase, and liquid helium for the last part of the cooling. Helium is used because it is one of the very few materials that remain liquid at these very low temperatures, and is also a superfluid: it can flow in a pipe without any mechanical friction, which is of course a very good property to avoid it warming up. The main drawback with the use of helium is its abundance (or rather lack of). Being lighter than air, there is basically no helium in the Earth's atmosphere (think about the helium balloons used for parties), and actually the name (from the Greek word for Sun, Helios) came from the fact that the first evidence for the existence of this element came by looking at the spectral lines of the Sun. Some helium gas can be found trapped in some rock formations, but it is still a rare and finite resource, and quite expensive. At CERN a large quantity of helium was bought before the start of the LHC, and the same helium is re-circulated in a closed circuit to keep the machine cold. Special pumps extract it from the magnets, cool it down again, and pump it back in, to ensure that each part of the accelerator stays at constant, and very low, temperature.

In addition to the dipole magnets, 392 quadrupole magnets are used to focus the beam. They are needed because the protons have all the same charge, and the electrostatic repulsion would tend to separate them; regularly spaced quadrupole magnets avoid the beam size to become too large.

Section 4.2: Acceleration and pre-acceleration

The accelerating cavities are the real engine of the accelerator. There electromagnetic waves with a frequency of about 400 MHz (in the range of radio waves, hence the name of "radiofrequency" cavities) give a push to the various particle bunches running in the ring. The proton beam, in fact, is not a continuous stream of particles, the protons are organised in about 1000 bunches (this number will more than double from 2015) moving in each direction, each of them having a length of a few cm and the thickness of a human hair. Each bunch is accelerated by the cavities by up to 5 MeV per metre, and only in a short region of the accelerator. This is possible because the protons almost do not lose any energy circulating in the ring. Moving at about the speed of light, the protons can cross the 27 kilometres about 11,000 times per second, and the acceleration phase lasts about 20 minutes; since in 20 minutes there are about 1000 seconds, it means that the protons cross the cavity region about 10 million times during the acceleration, and this is enough to bring the energy to the final value of 4 TeV (6.5, or perhaps 7 after 2015).

However, protons cannot enter the LHC at rest; they are not accelerated from zero energy — it would be like starting a car in fifth gear. The magnets, the cavities and the whole infrastructure of the LHC is optimised to work only in a given energy range. Protons must be pre-accelerated before entering the big machine, and this is done by re-using the accelerators that CERN has built in over 60 years of history. Machines that were once built to be the biggest and most powerful in their days, like the PS and the SPS, are now pre-injectors of a five-step accelerator chain: protons are extracted from a small tank of hydrogen gas, where they are stripped of their electrons using electrical fields; they are then sent to a linear accelerator (LINAC 2), followed by the first circular machine (the Booster, over 40 years old), the PS (the oldest accelerator of CERN still operational, which saw its first beams in 1959!), the SPS (in operation since 1976) and finally they enter the LHC through two transfer lines. Protons are injected from the SPS into the LHC with an energy of 450 GeV (about 500 times the value of their mass) and with a speed 99.9999% that of light. The final acceleration in the LHC increases the energy by a factor 20, but since no particle with mass can ever reach (let alone surpass) the speed of light, the margin of speed increase is very limited: the proton speed at the end of the acceleration will be 99.999999% of the speed of light — a very modest gain in percentage, but what really matters to produce new exotic particles is the increase in energy, because it is the beam energy and not the speed that is converted into mass.

The injection of the beam from the SPS can take about an hour, and the acceleration less than half an hour. In general the preparation phase to the collisions can last 4–5 hours, and the beams can collide for a total of 15–20 hours (the record so far is 26). However, as time goes by, the beam intensity decreases due to collisions between the beams and with residual gas in the beam pipe, so at some point it is not worth it any more to keep colliding, and it is better to extract the beam from the machine and inject a new, high-intensity one. The operation of extracting the proton beam from the accelerator is not trivial. The kinetic energy of the LHC beam is similar to that of a train running at 150 km/h, all concentrated in a radius going between 1 mm and 10 microns. This enormous energy density, if out of control, can destroy several magnets one after the other.

Section 4.3: Beam extraction

The most common problem of a cryogenic magnetic system is the magnet quench. A quench happens when a magnet loses superconductivity, typically due to the fact that even a very small region of the superconductor for some reason warms up and goes above the critical temperature. When this happens, 13,000 amperes of charge, used to run free in a superconductor, meet a resistance and warm up the magnet in a very fast way (systems are even in place to have a controlled rather than a disordered quenching, should that happen). The magnet loses its magnetic field very quickly, and the beam will feel in its trajectory a smaller total magnetic field. How small? Since there are about 1000 dipole magnets, the loss of one of them will result in an integral field smaller by 1 per thousand, which seems quite a small number; but if we consider that the radius of curvature of the beam is about 4 km, the protons would tend to go to an orbit that is larger than the standard one by 4 metres — clearly this is impossible, since they would first hit the walls of the beam pipe, destroying it.

To avoid this, a "safety exit" for the beam has been designed. A region of the accelerator has been dedicated to the "beam dump"; there, the beam passes close to a region with opposite magnetic field that would bend the protons outside of the accelerator instead of keeping them in the circular orbit. Whenever there is the need of stopping the experiment, or in any case of problem with any accelerator equipment, a magnet with very fast rising time deviates the beam towards this region of opposite field, and then out of the ring to impact a 7 metres long cylinder of graphite, surrounded by 750 tons of steel and concrete, where its energy is completely absorbed. The beam can be dumped by an automatic system in less than a millisecond, and it is very important that this safety system works at all times — it is too dangerous to keep this huge density of protons in a machine that's not perfectly working. It is better to dump a good beam because of a false alarm than to risk damaging the machine.

Section 4.4: Proton accelerator vs electron accelerator

Why do we collide two beams of protons?

First of all, out of all possible particles, only protons and electrons (plus their antimatter correspondents) are practical to store in a ring: neutral particles (like neutrons and neutrinos) do not feel the electromagnetic fields, so are impossible to accelerate with cavities and guide with magnets; all other particles are short-lived so it would be impossible to store them in a ring for hours. Protons and electrons have very different properties: electrons are elementary particles, so all their energy is available in the collision, and each collision will have exactly the same energy, and perfect energy and momentum conservation. Protons on the other hand are made of quarks and gluons, so every time two protons collide, in reality it is a quark or a gluon (at least at first order) of the first proton colliding with another quark or gluon of the second. Every collision will have a different effective energy, and momentum is only conserved on the transverse plane. Moreover, the quarks or gluons from the two protons that have not interacted will produce more particles in the detector, and become background noise to the main, hard collision processes.

From all these considerations, it looks like electron colliders would be superior to proton ones, and it would be true if the energy reach of the machines were the same — every physicist would love to have an electron–antielectron collider with the same energy as the LHC. We know that at CERN the LEP electron–antielectron collider has been running for over ten years in the tunnel now occupied by the LHC before it was dismantled. The reason for this choice is the reach in energy. Even if they produce much cleaner collisions than protons, electrons lose much more energy than protons when they circulate in a ring. Actually, every charged particle undergoing acceleration (even the centripetal acceleration that keeps them in a circular motion) emits electromagnetic radiation, and loses part of its energy. The energy loss depends strongly on the particle mass, and is minimal for heavy particles like protons, even at the LHC energies; for the much lighter electrons, instead, it is a major problem. During the last phase of LEP, electrons were losing a few percent of their energy each turn, making it impossible to further raise the energy. In fact, for the same length of accelerator LEP had an energy about 70 times smaller than the LHC; and even the SPS, with its 7 kilometres of length, can accelerate protons to energies larger than those of LEP. Typically in particle physics we observed an alternation between proton machines, which can go to higher energy and therefore possibly discover new particles, and electron machines, which have cleaner signatures and can study in details the properties of these particles.

Section 4.5: Characteristics of the LHC

After the LEP measurements, a new, high energy proton machine was needed. When the SPS was run in collider mode in the eighties, protons were circulating in one direction and antiprotons in the other. Since the two beams of particles with opposite charges have also opposite velocity, the same magnetic field could keep them in the same orbit, allowing the use of a single magnet and beam pipe for both beams. The problem with proton–antiproton machines is that while producing protons is straightforward, production of antiprotons is much more complicated. We need to first accelerate protons, send them against a target, then select with electromagnetic fields the very few antiprotons that emerge from the collision. It is a process with very low efficiency, and it is not possible to achieve very large intensities for antiproton beams. For this reason it was decided to have the LHC as a proton–proton collider — that was the only way to have the large intensities needed to discover rare phenomena, but it has the large drawback that, having beams of particles with the same charge moving in opposite directions, two opposite values of the magnetic fields are needed, thus two independent beam lines for the two directions.

The beams of the LHC circulate in two separate beam pipes, separated by 19 cm, and each surrounded by its own magnetic system. There are enormous mechanical forces acting between the two magnets, therefore a strong mechanical structure made of steel and iron is needed just to keep the magnet together. The problem is that this heavy structure (weighting about two tons per metre) is in contact with the superconducting coil, so it has to be cooled down to a temperature close to that of the niobium-titanium coil. It is however not possible to freeze the whole tunnel, so the external part of the magnets is at room temperature. Between the inner mechanical structure and the outer magnet container there is a temperature gradient of 300 degrees, in just a few centimetres of distance. It is clear that thermal insulation is fundamental. A large fraction of the energy used to run the LHC goes into the cooling system, which keeps the central part of the magnets at a temperature of 1.9 absolute degrees. Insulation is performed through vacuum: air is sucked out of the membrane between the inner and outer parts of the magnet, but perfect vacuum does not exist, and in addition magnets need support to stay above ground. Each magnet stands on three pedestals, which are designed to minimise thermic exchange with the exterior, but there are still about 3 watts of power (less than a calorie per second, a power similar to that of a bicycle's lamp) that enter the magnet. Even if that is not much, at these low temperatures it is sufficient to warm up the magnet very quickly, and this is why we need a liquid helium pipe inside each magnet, all along the length of the accelerator.

In addition to the dipole magnets, which constitute 80% of the length of the accelerator, the LHC has 392 quadrupole magnets. About half the size of the dipoles, the quadrupoles can be compared to the converging lenses of an optical system: they are used to focus the beam made of protons with identical charge, which tends to become wider after each turn. The quadrupoles, which as the name indicates are made of four coils, produce a magnetic field shaped like a cross, which can compress the proton bunches along the horizontal or vertical direction. Usually in the LHC there is a sequence of three dipoles, a vertical quadrupole, another three dipoles, a horizontal quadrupole, and so on. This helps to keep the transverse beam size constant, while higher-order magnets (sextupoles, octupoles and decapoles) provide additional corrections to the beam trajectory and the final focusing of the proton bunches before the collision.

With over 1600 big magnets, there must be an equivalent number of junctions — places where a magnet is connected to the next. This is one of the most delicate parts of the accelerator, since each magnet shrinks by several centimetres when cooled from room temperature to the operational temperature of −271°C. Overall, the whole accelerator shrinks by about 80 metres, but it cannot move. For that, the interconnections between the magnets have moving parts, made like an accordion, which compensate for the change in size of the magnets. Note that during this change, all electrical contacts, as well as the vacuum in the proton tube, have to be preserved.

As explained previously, the real engines of the accelerator are the cavities, since the constant magnetic fields of the dipoles and the quadrupoles cannot give energy to the particles, but only change their trajectory. The strange thing of the LHC is that we only have 16 cavities, 8 per beam, arranged in groups of four. Overall, a very small fraction of the length of the LHC is a real accelerator!

As it was said, this is possible thanks to the fact that protons almost lose no energy when circulating, and to the large speed that makes the protons cross the cavity region about 10 million times for a 20-minute long acceleration session. Having more cavities would only have shortened the acceleration time, but this would have been at the price of fewer magnets, so smaller integral of the magnetic field and eventually smaller maximal energy achievable in the collisions.

But how do these cavities work? The protons are not a continuous flux, but are grouped into bunches, each of them with about 100 billion protons. During Run 1, there were about 1000 bunches circulating in each direction, while in Run 2 this number will be more than doubled. Naively, one would accelerate a positive bunch with a negative charge, but when the bunch goes to the other side of the charge it would still be attracted by it, therefore slowing down. To have a net acceleration we need an oscillating charge that becomes positive, therefore pushing the protons, after the bunch has passed. But rather than having charges, what really matters is an alternating electric field. This field is produced outside of the cavities with instruments called klystrons (like the antennas used in microwave ovens), and transferred using wave guides to the centre of the cavity. The field is in phase with the arrival of the various bunches, such that, even though the field in individual cavity constantly oscillates with a frequency of 400 MHz (in the region of the radio waves, so the cavities are called "radiofrequency"), if we follow the motion of a bunch it will always be in a region of positive field, which pushes it.

A parallel can be made with a surfer riding a wave: even though a wave is an oscillation of the sea surface (it is enough to look at a buoy in the middle of the sea, moving up and down as the waves move), the surfer will always be in phase with the wave, always having the crest behind him, and can get energy from the wave.

In the accelerator we have the additional complication that particles are also accelerated, so the frequency of the electromagnetic wave has to increase to account for the shorter time it takes to move from one cavity to the next; however, since as we have seen the relative acceleration of the protons is very small, being these particles already very close to the speed of light as they enter the accelerator, in practice the oscillating frequency only has to increase by a few hertz. It is anyway important that this change occurs, since otherwise particles would get out of phase with the electromagnetic wave and cannot be accelerated any more.

Figure 4:   The tunnel of LHC with a dipole magnet. The internal structure and the beam pipes are clearly visible.

Chapter 5

Particle Detectors in the LHC

While the LHC machine can accelerate protons at energies over 7000 times the value of their mass, and collide them creating about 40 million "small Big Bangs" at four collision points, it tells us nothing about what are the results of these collisions. For that, huge particle detectors are placed around the collision points: very large and sophisticated instruments measuring the results of these collisions with an extreme accuracy.

When two protons collide, their kinetic energy gets transformed into mass, through Einstein's equation $E = mc^2$.

New, heavy particles are created, which live a tiny fraction of a second, and then decay into other, more stable particles. The heavy states created in the collision are not directly observable, but they have to be reconstructed from their decay products. The detectors have to be as hermetic as possible, to be able to measure all decay products of these heavy states, very quickly and precisely.

The four detectors placed around the LHC collision points are called ATLAS, CMS, LHCb and ALICE. These enormous instruments (ATLAS, the largest particle detector ever built, is roughly a cylinder with a length of 44 metres and a diameter of 25) are sometimes compared to gigantic digital cameras, since it is indeed possible to have visual images of the collisions. However, a camera takes pictures of objects outside of it, while the sensitive parts of the detector are crossed by particles from the inside. Moreover no camera would ever be able to shoot 40 million pictures per second!

In the past, real film cameras were used to observe the tracks left by the particles in detectors called cloud chambers or bubble chambers — charged particles would produce condensation in a thick cloud, or little bubbles by locally warming up a metastable liquefied noble gas.

But even if the beauty of the pictures taken by these detectors is still unsurpassed, data analysis had to be performed by hand, and rates were very low — in just a few seconds, the LHC detectors can collect as much data as that taken by all these optical detectors in every laboratory in the world, in their whole history.

Even if the modern LHC detectors can be read out every 25 nanoseconds, and therefore collect up to 40 million events per second, the storage capacity allows to record for later analysis only a very small fraction of this rate. Each collision is stored in a file with the size of a few megabytes, roughly the same as a high-resolution photo. And there is no place in the world where it is possible to store 40 million pictures per second, and do that for more than 10 years. The maximal rate that can be stored by a single detector is about 1000 events per second, and it already represents roughly the amount of data exchanged by all telecommunications in the world during that second. But even though it is a lot, this means that only a very small fraction of the number of collisions that the accelerator provides can be stored. Clearly the goal is to store the collisions that are considered the most interesting. An online system, called trigger, performs a rough real-time analysis of all collisions, classifying them into various categories (for instance in ATLAS there are over 1000 different event categories), according to the kind of particles detected and their energy.

Each category gets a weight that depends on its physics interest (and also on how common that category is). Some very rare events are considered top class and are recorded without any scaling factor; while for the vast majority, only a fixed fraction, sometimes very small, of the events in that category is stored.

The inverse of this fraction, called prescale factor, can be as large as a few billions (for very common events, where low-energy hadrons are produced, a topology with a rate of hundreds of megahertz, but of which only a few hertz are recorded), down to ten or less for rare events that are considered to be interesting. The final choice of the prescale factor is the subject of many discussions among the physicists working in the trigger, since opinions of which kind of events are to be collected in larger quantities may differ, and of course there is always the danger of recording a large number of interesting but not essential data, while missing potential discoveries because the importance of some categories has not been properly recognised. The physicists involved in the trigger system are working very hard to minimise these risks by constantly refining the algorithms and the criteria, and using "random" and "unbiased" triggers that also select events with no specific requirement.

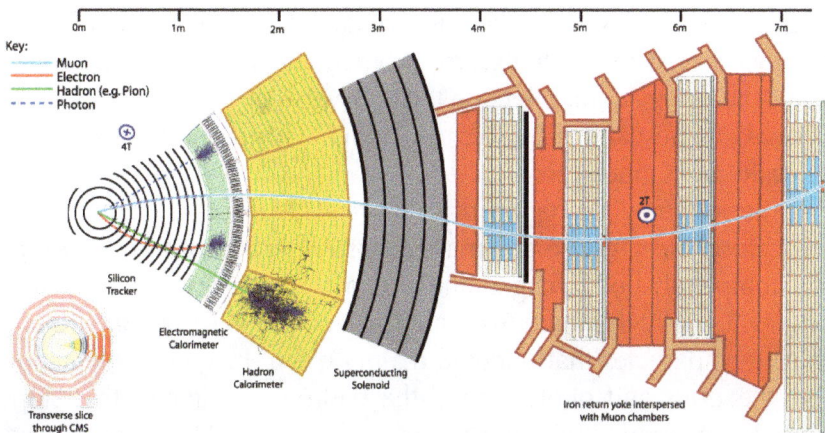

Figure 5: Different kinds of particles measured by CMS. The first detector after the collision point is a high-precision tracker; then electrons and photons are absorbed by the electromagnetic calorimeter while hadrons are absorbed by the hadron calorimeter. Muons are the only observable particles that can go through, and are measured by the external muon chambers.

But how do particle detectors measure the particles produced in the collision? Most of modern collider detectors (with the notable exception of LHCb) are approximately cylinders, with a horizontal axis corresponding to the axis of the accelerator that enters the detector from both sides. Sub-detectors also have a cylindrical geometry, and are located one after the other like the skins of an onion.

First, the trajectory of particles is measured in a low-density region, made of several concentric cylinders of very thin silicon, the same material used to make the chips of computers. The silicon allows measuring the position where charged particles cross each cylinder with a precision of a few tens of microns; a special "pattern recognition" software joins the dots and reconstructs the trajectories of the particles. Typically, the tracker is also immersed in a strong magnetic field with the axis parallel to that of the cylinders, such that the particles emitted in the collision undergo a helicoidal trajectory due to the magnetic force (the so-called Lorentz force). Positive and negative particles can be distinguished because they turn in two opposite directions, and the radius of curvature of these trajectories is proportional to the transverse momentum of the particles (basically, the energy component on the transverse plane). In other words, very energetic particles go almost straight, while the less energetic ones are bent much more.

The intermediate region of cylindrical detectors is dedicated to the calorimeters: dense and massive instruments that aim at absorbing completely both charged and neutral particles and measure their energy. Usually there are two calorimeters: the electromagnetic one that stops particles that undergo mainly electromagnetic interaction like electrons and photons, and the hadron calorimeter that stops particles interacting with the strong nuclear force, like protons, neutrons, pions, etc. Even though it is not possible to distinguish the particles from their energy deposition in the calorimeters, it is at least possible to distinguish charged and neutral particles from the presence of a track in the inner silicon detector, where only charged particles leave a signal. So, a deposition in the electromagnetic calorimeter is very likely to be an electron if a charged track is pointing to it, a photon if no nearby track is found.

Only two kinds of particles are not absorbed by the electromagnetic nor by the hadron calorimeter, and they are the muons and the neutrinos.

Muons are like heavy electrons — charged particles that are therefore measured by the inner silicon tracker, but which due to their mass are not stopped by the electromagnetic nor by the hadron calorimeter. They are however measured by the external muon chambers — very large position detectors (similar to the inner tracking chambers, but much bigger and made of cheaper technologies than silicon) placed outside the calorimeters. Any charged particle would leave a signal in the muon chamber, but muons are the only ones that can get there. Their energy is measured by their curvature in the magnetic field which is created in the external part of the detector.

The other particle that does not get stopped by the calorimeters is the neutrino. Neutrinos are neutral, so are not seen by tracking chambers nor by the external muon chambers, and are not stopped by the calorimeters, so they are completely invisible even in large detectors like those of the LHC. It is however possible to identify the production of a neutrino in an indirect way using a very hermetic detector. We know that energy and momentum are conserved in these collisions, so the final-state energy should be the same as the energy of the colliding protons, and the total momentum should be zero. Since it is not possible to identify particles emitted down the beam pipe, the energy conservation cannot be used, and not even the projection of the momentum along the beam axis. It is however possible to use the transverse momentum conservation: an event should be balanced if looked in the transverse plane. When this does not happen, so when momentum emitted in one direction is not compensated by an equivalent amount in the opposite direction, the most likely explanation is that an undetected particle, like the neutrino, has been produced. It is therefore possible to measure the direction and transverse momentum of a neutrino by measuring everything else, and this measurement can be quite accurate, considering that we are dealing with a particle that cannot be measured.

While these were the general characteristics of all collider detectors, each of the four LHC experiments, placed around the collision points of the proton beams, has its own specific characteristics.

Why four detectors? First, the experimental method requires reproducibility of the results, especially given the complexity of the collisions, and the statistical nature of the processes under investigation. It was therefore mandatory to have at least two general-purpose detectors that are fast enough to be read-out at the very high rate of the LHC, and as hermetic as possible to be able to reconstruct most of the collision. ATLAS, just in front of the main CERN site, and CMS, located at the diametrically opposite side of the ring, at about half an hour driving distance, are the two general-purpose detectors. They have been built using quite different design philosophies and technologies by two completely independent collaborations, to minimise the possibility that an effect independently observed by both detectors would be an instrumental fake. To further reduce the possibility of mutual influence, and to guarantee the scientific integrity of the results by minimising rumours and preliminary results, the members of the collaborations are not allowed to discuss, even informally, results that have not been published or made public in the form of conference contributions. This measure may sound harsh, but it is important to guarantee the full independence of the results.

The other two detectors are specialised to specific topics.

LHCb can measure particles emitted at very small angle with respect to the beam direction, and can do that with very high accuracy. The geometry of LHCb is very different from that of the standard collider detector: while the others are approximately cylinders fully embracing the collision point, LHCb has the shape of a pyramid with a horizontal axis. Moreover, the collision point is at the apex of the pyramid, so basically outside of the detector. This means that the majority of the particles produced in the collision (for instance, those emitted perpendicular to the beam axis) are falling outside of the LHCb volume and not measured. So a full reconstruction of each collision is not possible, but this is not needed to accomplish the physics goals of the experiment: measure particles going in one direction with the highest possible accuracy. As the name LHCb suggests, this experiment aims at measuring with the highest accuracy particles containing the "bottom" quark, also called the b-quark. These particles have a very short lifetime (of the order of picoseconds), and they decay in a collimated jet of hadrons, usually emitted in the forward region; this is why this detector concentrated all its efforts in reconstructing very well a small angle around the beam axis.

The geometry of the ALICE detector is again a cylinder, but this detector uses in its central tracking chamber a detector with gas instead of silicon; it allows measuring about one hundred points along the track trajectory, compared to about 10 for silicon detectors. Gas trackers are on the other hand much slower than silicon ones, and certainly are not able to take data with a rate of 40 MHz. ALICE is specialized for taking data during the special runs of the LHC (usually a month per year, each autumn) when lead nuclei are collided in the LHC instead of protons. The energy and density of these collisions is much higher than those of protons, but their intensity is smaller, and collisions happen every few milliseconds instead of nanoseconds. The results of these collisions have a very high particle density, and many points are required to properly reconstruct these tracks. This is why ALICE is a dedicated detector mainly used to measure these very dense heavy-ion collision environments.

Let us now discuss in more details the characteristics and components of the four detectors.

Section 5.1: ATLAS

ATLAS is the largest particle detector ever built. It has a length of 44 metres and a height of 25, and many components. Its name is short for A Toroidal Lhc ApparatuS, stressing the fact that the geometry of the external magnetic field is toroidal, meaning that the field lines are circular, surrounding the external muon chambers.

The central tracker is made of silicon detectors, where the position of charged particles is measured by the energy they leave in electrodes. Towards the centre of the detector, the electrodes have a spherical shape, allowing the reconstruction of all three coordinates

Figure 6: The ATLAS detector. This picture shows the end-cap of the calorimeter, surrounded by the eight external magnets (the grey tubes with the orange strips) and the muon chambers.

of the passing particle; more far away the electrodes are strips, which only allow the reconstruction of two dimensions, with the possibility of inferring the third coordinate by combining the information of several strips tilted by small angles.

Having three-dimensional information is very important in the central part where the particle density is very high, but it requires a very large number of channels of readout electronics; this is why in the larger external part, where it is easier to associate energy depositions in successive layers, the strip detectors are used. The precision of these detectors is remarkable: about a twentieth of a millimetre for the three-dimensional ones, and a hundredth of a millimetre for the two-dimensional ones.

A third tracking detector, made of narrow straw tubes, is placed outside of the silicon strips. Its space resolution is worse than that of the silicon, but the strength of the signal allows separation of the signal left by electrons from that of pions and other particles.

All of the inner tracker is located in a 2 tesla magnetic field, used to bend the trajectories of charged particles and measure their charge and transverse momentum.

The electromagnetic calorimeter is made of a lead absorber, with the shape of an accordion, immersed in liquefied argon. Electromagnetic particles produce showers of secondary electrons, positrons and photons in the lead–argon sandwich; and crossing the argon they produce ionisation electrons that are measured by electrodes.

The next stage is the hadron calorimeter, made of alternating copper tiles, used as an absorber, and scintillating plastic, which produces light when crossed by charged particles. The light produced this way will be proportional to the energy of the incident particle, and optical fibres will carry this light to photomultipliers, where it is converted into electric current proportional to the light and therefore to the particle energy.

The defining characteristics of ATLAS are its huge muon chambers, the largest sub-detector ever built. Hundreds of square metres of tubes containing gas and an electrical wire in the centre measure the position of the muons, the only detectable particles that can escape the calorimeters.

Muons are bent by the toroidal magnetic field, produced by eight large superconducting magnets, complemented by two additional magnets in the forward and backward regions. So ATLAS has a total of 11 large magnets: an inner solenoid, eight producing the central toroidal field, and two producing the forward–backward one. Knowing the position of each chamber and of the magnets is the main limiting factor to the precision of the measurement; the alignment is made possible by a real-time laser system that provides a precision of a twentieth of a millimetre over lengths of several tens of metres.

Being so big, ATLAS has been assembled entirely inside the cavern. During ten years, the various institutions (universities and research institutes member) of the collaboration (about 200) have built the various parts of the detector and shipped them to CERN, where the excavation of the huge cavern was taking place at the depth of 80 metres. Once the cavern was completed in 2004, the components of this detector were assembled over four years like a gigantic ship in a bottle.

Section 5.2: CMS

Figure 7: The magnet and central muon chambers of CMS as they complete their descent into the cavern. The calorimeter and inner detector will be later installed inside the big cylinder.

CMS stands for "Compact Muon Solenoid"; it is compact with respect to ATLAS, since with about 15 metres of height and 22 metres of length it is still as high as a 4-storey building. The other difference with respect to ATLAS is also suggested by the name: the magnetic field is a single solenoid, as opposed to ATLAS' multi-magnet toroidal system. Being smaller, it was possible to fully assemble and test the detector on the surface in a big ground-level hall. After the successful test, the detector was "sliced" into seven pieces, and each of them was very slowly lowered into the underground cavern using a gigantic crane; the various parts have been again assembled together and tested in their final location.

The magnetic field of CMS is produced by the largest superconducting magnet ever built. A cylinder with a length of 13 metres and a diameter of 6, its 4 tesla magnetic field over this huge size stores enough energy to melt 18 tonnes of gold. This single magnet produces the internal magnetic field used to bend the charged particles measured by the silicon detector, as well as the external field that bends the muons that escape the calorimeters. This external field is trapped by an amount of iron about twice that of the Eiffel Tower, and is opposite in direction with respect to that of the internal one. A muon that will be bent by the internal field first and by the external one after it exits the calorimeter will travel in an S-shaped trajectory, and be measured with a very good precision.

While the inner tracking chamber uses a silicon technology similar to that used by ATLAS (but for a total surface of about 200 square metres of silicon), the electromagnetic calorimeter is composed of nearly 80,000 crystals of lead tungstate (a combination of lead, tungsten and oxygen), each of which took about two days to grow, in two dedicated factories in Russia and China. This material was found after years of research and development, since it had to meet very stringent requirements: being as dense as lead to be able to stop very high-energy particles in a short distance, transparent like glass to allow the light produced by the particles to reach photo-detectors placed behind, and have its properties unchanged even after years of operations in a hostile environment. The advantage of crystal detectors is that being fully sensitive (no lead absorbers, for instance), their precision in measuring particle energies is very good provided each crystal is individually calibrated, a very complex task.

The hadron calorimeter is made of an alternating sequence of absorber material (consisting of over a million brass shell casing left over from World War II, courtesy of the Russian Navy) and wire chambers as sensitive detector. Both the electromagnetic and the hadron calorimeters are placed inside the big magnet: outside there is only the return yoke of the magnetic field, interleaved with the muon chambers. Here is another important difference between CMS and ATLAS: while in ATLAS the muons outside the hadron calorimeter travel in an air core, in CMS they cross the iron used to absorb the magnetic field outside the magnet. This allows a more compact design for CMS, at the expense of a slightly worse measurement of high-energy muons, whose trajectory can be deviated by their passage through the iron. Also, muons produced in the forward region are hardly bent by a solenoidal field, and the precision of their measurement in CMS is not so good.

The strong points of CMS are mainly its excellent tracker, as well as the crystal electromagnetic calorimeter.

Section 5.3: LHCb

Figure 8:   The LHCb detector in its cavern. The accelerator tube crosses the detector in the middle, with the collision point located at the right side of the cavern, inside the detector called "VELO".

LHCb is very different from any previous collider experiment. While the other detectors are approximately cylinders surrounding the collision point, LHCb has the approximate shape of a pyramid, with a horizontal axis, coincident to the beam pipe. The collision point is located slightly outside of the detector cavern, about at the apex of the pyramid. Particles produced in the collision, like those emitted at the centre of ATLAS and CMS, will go in all directions and most of them will not be measured by LHCb. Since in this detector only a small fraction of the particles produced are measured, a full reconstruction of the event is not possible, but is not even needed.

LHCb aims at reconstructing with very good precisions the decays of the fifth quark, the "bottom" or "b". While the top quark, the heaviest of them, has such a short lifetime that it will immediately decay without having time to form bound states with other quarks, the b can produce a large number of combinations with all other quarks, creating particles with lifetimes of the order of the picosecond. Since these particles are emitted at speeds close to that of light, it would correspond to travelling about a third of a millimetre, but here relativity plays an interesting game: like the cosmic muons, time for these particles is dilated with respect to that of the observers, by a factor equal to the ratio between the energy and the mass of the particle. For masses of about 5 GeV, a typical energy of 100 GeV results in an increase of the lifetime by a factor of 20, so decay length of 6 mm. Such distance is measurable, if very precise detectors are placed close to the collision point; silicon discs, roughly the size of a compact disk, placed vertically with the beam as an axis, will measure the position at which charged tracks cross them, with a precision of a few microns. This will allow a precision reconstruction of the particle trajectories, determining if they come from the main collision point or from a displaced secondary vertex, an indication that they are the decay product of another particle with a short lifetime. As long as all the decay products of the b hadron are contained inside the volume of LHCb (as it is usually the case), a full reconstruction of the decay can be performed, without the need of a full reconstruction of the rest of the event. Being dedicated at measuring just an angle of about 30 degrees around the beam axis, LHCb measures particle trajectories with very good precision, using silicon tracking devices and the particle momenta through their curvature in a magnetic field, provided by a large-opening magnet.

But measuring the trajectory and direction of particles cannot distinguish between them. Ideally, one would like to measure the particle mass, but a direct mass measurement is impossible at these energies; it is however possible to measure the speed of some particles using the Cerenkov effect. When a charged particle travels faster than light in a medium, it does emit a cone of electromagnetic radiation. We know that no massive particle can reach (let alone surpass) the speed of light in vacuum; however, light travels at lower speeds in a transparent material, and this is the working principle of lenses, or the reason why fish look bigger if seen through swimming goggles. If a muon travelling at 90% of the speed of light in vacuum enters a medium where light travels at 80% of its normal vacuum speed, the muon (which does not slow down much in crossing the material) will be faster than light in that material, and will emit a cone of light, equivalent to the "bang" emitted by planes flying faster than the speed of sound in air. The opening angle of this cone is connected to the speed of the particle, so a combination between this measurement and the momentum given by the curvature in a magnetic field allows an approximate measurement of the mass, sufficient to distinguish in most cases various kinds of particles. LHCb has two such Cherenkov detectors, while for reason of space and particle occupancy ATLAS and CMS are not equipped with them, so their ability to distinguish charged hadrons is much more limited.

To detect neutral particles, LHCb is also equipped with electromagnetic and hadron calorimeters, with the geometry of walls, and has a final array of muon chambers, where the muon momentum is measured by their curvature in a magnetic field.

One of the reasons why the physics of particles containing b-quark is so interesting is that for them the difference in weak interactions between matter and antimatter can be very large. Since we are made of matter, and looking at the sky no explosions due to annihilations between matter and antimatter are seen, we have to conclude that at least in the visible universe there is very little antimatter. The question why in the universe (at least the visible one) there seems to be much more matter than antimatter is a puzzling one, especially considering that the energy from the big bang should have created equal amounts of matter and antimatter. One of the possible solutions to this puzzle is that if the universe went through a phase of thermal non-equilibrium, and the laws of physics are different between matter and antimatter, the annihilation between matter and antimatter would be asymmetric — all antimatter could be annihilated and the current universe would be just the leftover matter. Weak interactions are the only phenomenon where different behaviours of matter and antimatter has been observed, and this fact has been studied in deep detail in a large number of experiments, LHCb being one of them. The picture that is emerging from these measurements is that this difference, which can be explained and understood in the Standard Model, is way too small to explain the current asymmetry of the universe. Even if this conclusion seems now quite established also thanks to the results of LHCb's first years of running, this experiment still has a lot to search for. In fact, any hint of new physics is likely to be first found in small deviations from the predicted behaviour of known processes. LHCb performs very accurate measurements of systems where the theory predictions can be very precise, so it is in a very good position to find these deviations, and therefore the first hints of the need to consider new theories to accommodate the new phenomena.

Section 5.4: ALICE

ALICE is externally more similar to ATLAS and CMS than LHCb, but is much "slower", being dedicated to the measurement of collisions between lead ions. Towards the end of each year the LHC is run in "heavy ion mode"; it means that instead of protons the machine accelerated nuclei of lead, each containing 82 protons and more than 120 neutrons. They reach a maximal energy of 2.76 TeV per nucleon, so the total energy of a nucleus is much larger than that of protons when the accelerator is running in normal proton mode. On the other hand, the collision rate is smaller: each bunch contains about 100 million ions (compared to the 100 billion of a proton bunch), and bunches collide 5 million times per second, as opposed to 40 million times. It means that the average separation between events is of the order of milliseconds, as opposed to nanoseconds for the proton run.

Figure 9: The ALICE detector, with the inner part visible through an open door of the magnet. This magnet is the same as the L3 detector, which between 1989 and 2000 measured the results of electron–positron collisions from the LEP accelerator.

A detector dedicated to the ion run like ALICE has less stringent requirements in terms of data acquisition speed. It allows the use of a tracking technology that was very common at the time of LEP, namely the time projection chamber. This device is a cylinder filled with gas, with a magnetic and an electric field parallel to its axis. Charged particles from the collision would ionise the gas, leaving a contrail of electrons, which are attracted by the electric field and will be measured by detectors placed at the base of the cylinder. This kind of detector is clearly slower and less precise than the thin silicon trackers used by the other experiments, but has the advantage that it can measure about one hundred points along the particle trajectory, as opposed to the ten points measured by the concentric cylinders of the silicon detectors. Reconstructing a track from many points has a decisive advantage in very dense final states like those produced in lead collisions, where several thousands of tracks are produced: it is impossible to properly match just ten points along a trajectory when many similar and very close tracks are also present. ALICE is complemented by a full calorimeter system and a forward spectrometer, which resembles in some parts the geometry of LHCb.

The main physics goal of ALICE is the measurement of the quark–gluon plasma. We know that now, due to the laws of strong interactions, quarks and gluons cannot exist in isolation, but are bound into hadrons, like for instance protons and neutrons. But there has been an early phase in the life of the universe when quarks and gluons were free, and actually the whole universe was a big plasma of these two particles. To reproduce these conditions, not only do we need high energies, but also high pressures and large nuclear media, a condition that is found in the lead–lead collisions of the LHC. The good tracking capabilities of ALICE, complemented with some particle identification capabilities, allow to experimentally demonstrate the production of this new state of matter, and study its characteristics.

ALICE is also taking data during normal proton runs, but the beam density is deliberately reduced close to its interaction point to have a rate in proton runs similar to that of the heavy ion one. Due to this reduced collision intensity, ALICE cannot measure rare events when running in proton collision mode.

Section 5.5: Data acquisition and analysis

The four particle detectors (plus other much smaller ones, not mentioned here) use quite different technologies to measure the particles emitted in the collisions, but they all produce electronic signals that are translated into numbers by specific converters, and ultimately become a stream of computer-generated data. The bunches cross 40 million times per second, and each time there are about 30 protons from each side colliding with 30 protons from the other, for a total of about a billion collisions per second. Since the detectors are anyway read out 40 million times per second, every collision will be the overlap of about 30 independent ones, and its interpretation is equivalent to watching a picture made of 30 exposures on top of one another!

Each of these snapshots can be written on a file with size of a few megabytes, the equivalent of a high-resolution photograph. Even if the detectors are read out at this rate, all this data cannot be stored. The maximal rate at which collisions can be practically stored is about 1000 per second. The goal is therefore to select the thousand most interesting collisions out of the 40 million, and this is the task of the so-called trigger system. This very important device, made of super-fast electronics and large computer farms, classifies all collisions in various categories (over 1000 per experiment), and each category will have a different priority level.

Only selecting the "interesting" events is a potentially dangerous approach, because if the priorities are set wrong, we risk rejecting very interesting collisions that could for instance lead to an unexpected discovery while filling the disks with more predictable but easily recognisable data. Clearly this risk exists, but in the design of trigger criteria the physicists try to be as model-independent as possible, and there is even a random trigger that (even with enormous prescale factors) selects events without any special requirement. In any case, given the technological limitations of the data-acquisition system, the only way to store everything would have been to reduce enormously the intensity of the accelerator, making it impossible to observe very rare events like the production of the Higgs boson or (hopefully in the future) of supersymmetric particles.

Once a specific event is selected by the trigger, it is stored on tape (which is still the cheapest storage technique for large data samples). The information recorded is very much low-level, like the current in a specific channel of the calorimeter, or the charge of a capacitor. Reconstruction and calibration codes convert this low-level information into something with more physical meaning, like the fact that a specific particle is produced in a given direction and with a specific energy.

These processed data are smaller in size than the full raw data, and are used by the physicists for the analysis. Anything that goes from these big files and a scientific publication ready to be sent to the journal is the proper analysis work, the activity that occupies most of the physicist's time once the detector starts taking data. Analysing the LHC data is clearly not an easy task, and is performed in teams, comprising from a few members to hundreds, according to the complexity of the topic. Apart from the core members of the team analysing a specific physical process, many other people are involved, from those helping on the data taking, those working on the trigger system, on detector calibration, the internal review team etc. We can say that each analysis is the overall work of the whole collaboration, and in fact each scientific article is signed in strict alphabetical order by all members of the experiment, a number that for ATLAS and CMS reaches 3000 physicists. Sometimes the author list is longer than the paper itself! It may seem strange to publish articles this way, but it is really difficult in such a big collaboration to understand who gave an even minimal contribution to each result, and having the whole experiment sign all papers increases the collaborative spirit. Clearly the number of published articles, usually an important measure to evaluate the productivity of a scientist, cannot be used in experimental particle physics, since it just corresponds to how long a scientist has been a member of a collaboration, and it is not rare to have scientists being authors of hundreds of articles, even if major contributions have been given only to a much smaller number; there are however other ways to assess the participation to the analysis, like internal notes (only signed by the members of the analysis team), or conference presentations. An article can present the measurement of a specific quantity or property, or the results of a search for new phenomena (or both); usually for measurements specific "unfolding" techniques are used to remove from the measured data the detector effects, and present a result that could be directly compared to those of other detectors or to theoretical predictions. Searches can either look for a specific new particle, or model, or be more general and model-independent, presenting the results in terms of deviation of parameters from the expected values.

The results of searches are usually limits in a complex multidimensional parameter space: it is in fact never possible to completely exclude the existence of a particle, but at most it can be stated that if this particle exists its mass or coupling has to be larger (or smaller) than a given value. Both the unfolding technique used in measurement or the limit-setting strategies used can require very involved statistical techniques, and each particle physicist needs to be quite an expert in statistics; there are in each experiment special groups dedicated to studying in detail complex statistical problems, and provide recommendations for the most difficult cases.

No modern data analysis could be done without the help of Monte Carlo simulations of the detector and of the underlying physics processes. The Monte Carlo method is so-called because it makes vast use of random numbers (like the results of the roulette in the famous casino) to simulate physical processes according to specific distributions. In quantum mechanics it is not possible to exactly predict the outcome of an experiment, but only the probabilities of specific results; also, apparently identical proton collisions always produce different final states. A large number of collisions are simulated using random numbers distributed according to distributions derived from theory; for each of these collisions also simulated is the interaction with the detector (described in a very complex computer code with the accuracy of each single screw), and detector response files, identical to those coming from real data but entirely simulated using computers, can be produced. The difference with respect to data is that in this case the simulated process is entirely known, as are the characteristics of the particles before and after interacting with the detector (the difference allows to study the effects of the detector on the measurement). Knowing the effects on the detector of each physical process independently helps designing the event selection and separate signal events (those that the specific analysis is interested in finding) to the background (collisions giving similar final states in the detector, but coming from other underlying processes). Monte Carlo generators have come now to a great level of sophistication, and their development is the consequence of the work of many theorists over several years.

Most of the heavy states produced in a collision have a very short lifetime, and decay before being measured in the detector. So, how is it possible to be sure of their production, and measure their properties?

A very important quantity exists in physics — the invariant mass — that is conserved throughout a decay. So if a heavy particle decays into a series of secondaries, it is possible to calculate from the energy and direction of these secondaries their invariant mass, which will be equal to the mass of the heavy state before the decay. So to look for a new particle (and this is precisely the method followed for the Higgs discovery, as we will see later), first a theoretical hypothesis is made on the way this particle can decay. Then the collisions where this decay mode is present are selected among the billions of others. But observing them is not enough — often other background processes produce exactly the same final state as the object of the analysis. The production of a particle is indicated by a peak in the invariant mass distribution of many events from this final state, while background events usually do not show any structure in the invariant mass distribution.

Analysing the LHC data requires an enormous amount of human effort, but also a large computing power, both in terms of processing and storage capacity. Physicists write computer code that is executed over billions of events, and modifications and improvement in the analysis strategies require this rerun to happen several times.

Until the early '90s, computing power for particle physics experiments was provided by large "mainframe" supercomputers, a bit like those seen in science-fiction movies. Since more than 20 years, it is possible to obtain even better performance by connecting between them a large number of computer, each individually less powerful than the supercomputer, but also much cheaper and more flexible from the point of view of maintenance and upgrades. Slowly CERN's computing centre transformed: from hosting a small number of big units, it got filled with racks containing hundreds or thousands of workstations, computers larger than a PC, but much smaller than the old mainframes. With the spread of the Linux operating system, it was finally possible to use on standard PC's the same software used by the workstations and by the mainframes. The computing centre changed again: it now contains over 10,000 PC's, like those everybody could buy in a shop (but with very reliable components, otherwise having so many the rate of failure would be almost daily), all connected between them with high-speed links, providing computing and storage power that would never be achievable with a single unit. The computing power installed in CERN's computing centre is still not sufficient to analyse the large amount of data from the LHC. And it would not even be appropriate, given that the physicists are in universities all over the world, to have it all concentrated in Geneva. The concept of distributed computing has been moved to a higher level by the use of the computing grid. The idea is to connect over 200 computing centres all over the world, for a total of 100,000 processors, and use this whole system to analyse the LHC data.

These computing centres, installed in universities and research centres all over the world, are organised in a hierarchical structure, with the main controls being at CERN (the so-called Tier-0), then 12 Tier-1 centres (approximately one per big country or geographical region), and many smaller Tier-2 centres located in universities. The Tier-3 is the computer of the final user, namely the physicist doing the analysis. After having tested his analysis code on a small fraction of data (this testing can nowadays be done even on a laptop), the physicist willing to analyse a large data set sends his code "to the Grid"; it means that the Grid infrastructure will look around the world for idle computers on which the analysis code could run. The job is split into sub-tasks (typically, the same code will run on various blocks of data in parallel and independently), and after the run is finished the final result is again put together and sent back to the user, who does not even need to know where physically his code was executed. Obviously this concept that optimises the resources on a worldwide scale can only work if several copies of the data are present; there would be no advantage in using a processor located in Japan to access data in Geneva: it would create an enormous internet traffic, without real performance gains. So, after a first reconstruction stage, the LHC data is copied to various locations, to allow many physicists from various geographical regions to access it. The Grid is in a sense the evolution of the web; while the latter allows computers from all over the world to exchange information, the Grid allows the exchange of computing power. And like the web, this idea is spreading outside of particle physics: computing grids, using different hardware, but the software developed at CERN for the LHC, are now used in other areas of science where international collaborations have large computing needs. Climate science, astrophysics, disaster prevention, and molecular biology are some examples that now use grids for their computing infrastructure.

Chapter 6

Main Physics Topics
at the LHC

Let us now look in some detail at the main physics topics that are studied at the LHC.

Section 6.1: QCD and the Standard Model

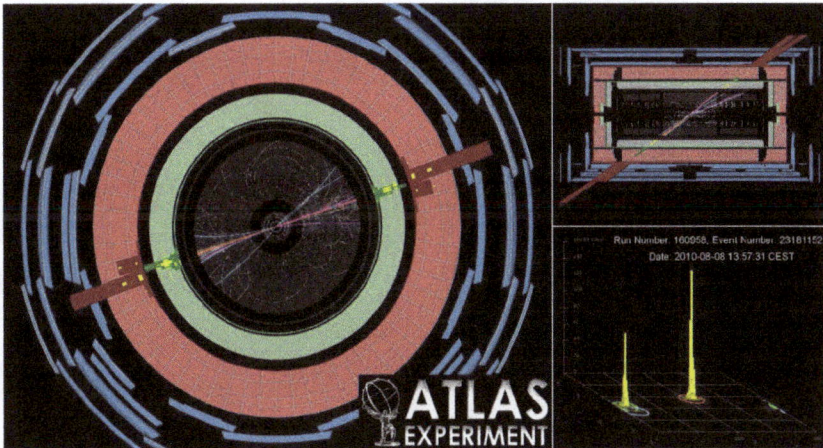

Figure 10: A collision in ATLAS showing two high-energy jets.

Jets are collimated sprays of hadrons, produced by a quark or a gluon. Emissions of quarks or gluons, apart from being the most likely process at the LHC, can occur in association with many other particles, so the study of jets and their properties is fundamental for each physics topics, and measuring the probability of jet production is one of the first measurements performed at the startup of the machine. Jets can also come from the hadronic decay of heavy particles, like the top quark (which can decay into three jets) or the W, Z or Higgs bosons (which produce two jets). When the parent particle is very energetic, the jets produced are very collimated and close to each other, so they will be reconstructed as a single jet in the detector. Specific techniques have been developed to analyse the internal

CMS Experiment at LHC, CERN
Run 133877, Event 28405693
Lumi section: 387
Sat Apr 24 2010, 14:00:54 CEST

Electrons p_T = 34.0, 31.9 GeV/c
Inv. mass = 91.2 GeV/c^2

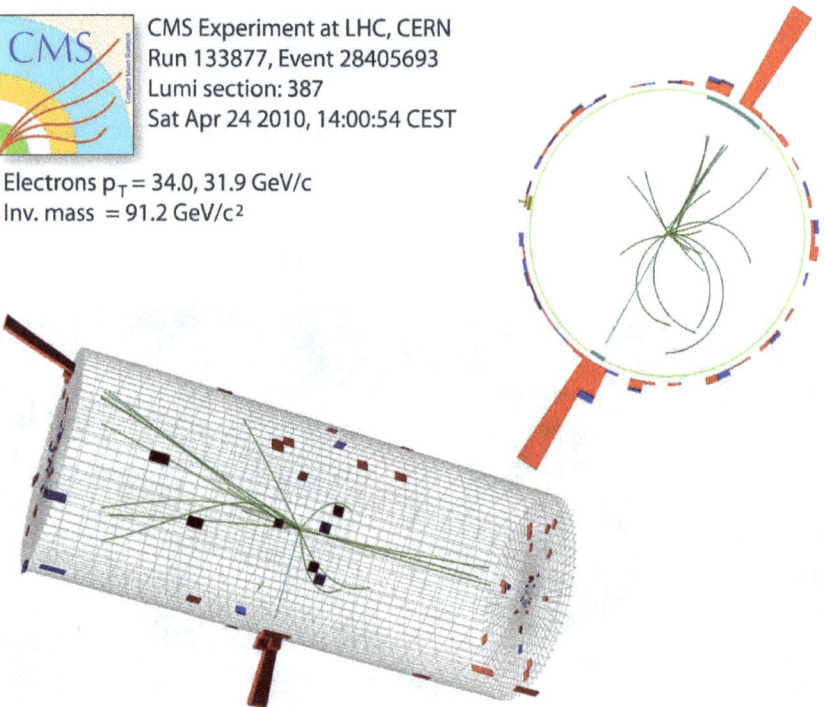

Figure 11:   A collision in CMS where an electron and an antielectron are produced. They are probably coming from the decay of a Z boson.

structure of these large jets, and tell if they are likely to come from the decay of heavy states.

Another important measurement is the production of W and Z bosons, the "vectors" of the electroweak force. The properties of these particles have been studied in great detail in the predecessor of the LHC, the LEP collider that smashed electrons and positrons instead of protons. Still, it is important to measure the probability of producing these particles in a proton machine, especially in association with jets. Also production of pairs or even triplets of vector bosons can be studied at the LHC much better than at LEP, providing measurements of fundamental couplings (and search for new physics). Finally, it is expected that after several years of running the LHC could reach an unprecedented precision in the measurement of the W boson's mass, one of the fundamental parameters of the Standard Model.

Section 6.2: Physics of the b-quark

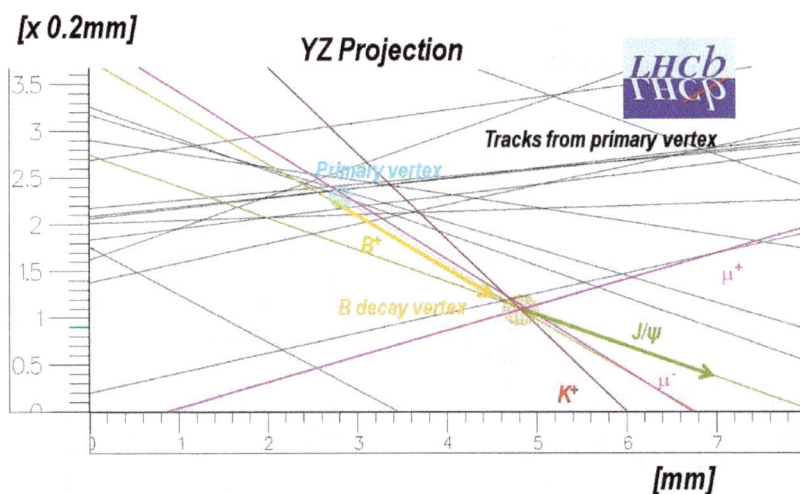

Figure 12: In this collision recorded by LHCb, a B+, produced in the primary vertex, decays after about 2 mm into a J/Ψ particle (which in turns produces a pair of muons, µ+ and µ-) and a kaon, indicated as K+.

The b-quark can mix with all other lighter ones, and produce a large quantity of hadrons. These particles decay with lifetimes of the order of picoseconds, but as we have seen, at LHC energies the time dilatation due to relativity can make them travel a few mm before they decay, allowing reconstruction of a secondary vertex in the detector. While LHCb is specialised in the reconstruction of particles with a b-quark, ATLAS and CMS also have dedicated groups studying these particles. While they have worse reconstruction capabilities than the dedicated detector, being faster ATLAS and CMS can record a larger number of collisions, making them superior in the search for very rare decay modes, which if enhanced with respect to the predictions can indicate presence of new physics.

Section 6.3: The top quark

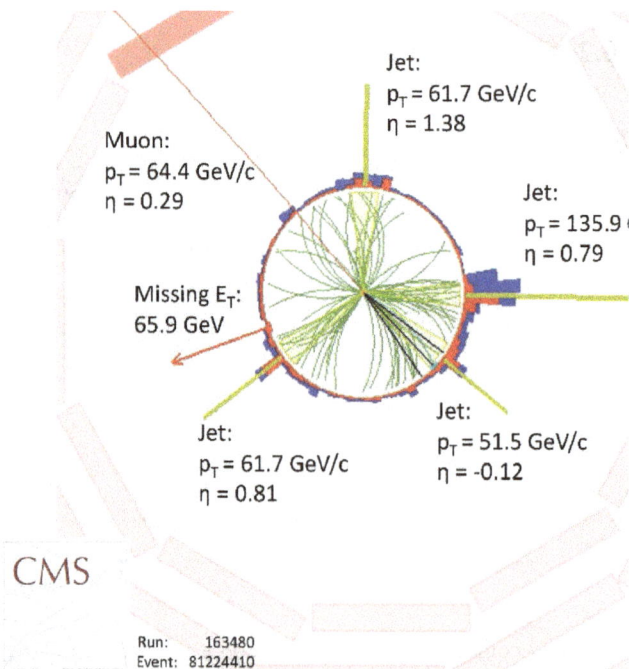

Figure 13: A collision in CMS, where each of two top quarks decays into a b-quark and a W boson. One of the two W's decays into two jets, the other into a muon and a neutrino, indirectly observed as missing transverse energy, "Missing E_T".

The top quark, heaviest of them all and the last one having been discovered, has a mass about 200 times that of a proton. It decays before having the time to form bound states, so no hadrons containing it exist, and the bare properties of this quark can be studied. The mass, its production probability and decay modes have been studied with great precision, thanks to the large number of top quarks produced. In almost all cases, a top would decay into a b-quark, which will create a jet starting from a displaced vertex, and a W boson. The W can either decay into a pair of jets, or into a charged lepton and a neutrino, and depending on the two decay modes, events can be classified into hadronic, semileptonic and fully leptonic.

Many models of new physics predict that new particles (which are probably heavier than all known ones, otherwise they would have been discovered already!) couple with the top more than with the other quarks, so deviations from the expected behaviour are more likely to happen here than elsewhere.

Section 6.4: Search for new physics: supersymmetry

There are several models inspired by the supersymmetric principle, namely the existence of a boson for each know fermion and a fermion for each known boson. These new particles can also mix between them, producing a large number of states; since they have not been discovered, supersymmetric particles must be heavier than their Standard Model correspondents. These theories are very popular since they would solve one of the theoretical problems of the Standard Model, the fact that the Higgs mass is close to the mass of the W and Z bosons, while if, as expected, the Standard Model is embedded in a more complete theory, its natural mass should be much higher. Supersymmetry, through cancellations at quantum level, provides a mechanism to stabilise the Higgs mass and keep it small. It also predicts the existence of particles that are weakly interactive and massive, ideal candidates for dark matter. Since for most theories supersymmetric particles can only decay into other supersymmetric ones, if they are produced at the LHC they would decay in a chain that terminates with the lightest supersymmetric particle, which cannot decay any more, but is stable and not detected, just like a neutrino. An experimental signature for the existence of SUSY would be an excess of collisions where many ordinary particles like leptons and jets are present, as well as large missing transverse energy, in excess to what is predicted by processes that have final state neutrinos like top or W production.

Despite many searches in previous colliders and at the beginning of the LHC run, no trace of supersymmetry has been found. If not found at the LHC, it is unlikely that this theory could perform the task of stabilising the mass of the Higgs, so it would lose part of its appeal.

Section 6.5: More searches for new physics

Many other theories have been developed to complete the Standard Model. These include the existence of smaller constituents than quarks, of heavy copies of the Z and W bosons, of next-generation leptons or neutrinos. Several theories also postulate a "hidden sector" with new set of particles and laws of physics, very weakly connected with the particles we know, so very difficult to observe.

In general, while for a given SUSY model it is possible to derive quite precise predictions, the search for generic new physics is usually more model-independent. Several basic distributions are measured, and deviations from the expected behaviour are searched for. When these are not observed, the result can be interpreted as limits over a large set of models. For instance, a new resonance could produce two jets. A classic generic search for new physics consists of taking a distribution of the invariant mass of all dijet events, and look for the presence of peak, indicating the decay of a new particle. If this peak is not observed, limits are placed on a variety of models, since each would predict a different behaviour of the invariant mass distribution. So a single measurement can have several interpretations, and help constraining several new physics models.

Section 6.6: Black holes?

There has been quite a lot of media buzz, at the start of the LHC in 2008, about the possibility of producing black holes from proton or

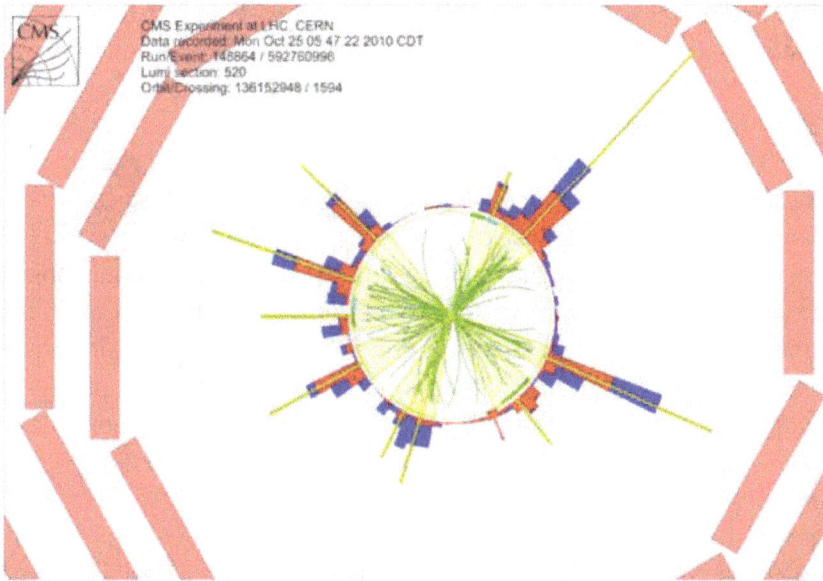

Figure 14: A collision showing a large number of jets, recorded by CMS. An excess of these kinds of events with respect to the expected background could signal production of mini black holes and their successive evaporation.

heavy nuclei collisions. CERN scientists did not seem very worried by this issue, but the laboratory did anyway ask some authoritative theorists to investigate scientifically the issue, and to publish a very detailed scientific article on the subject. The point is that some very exotic theories predict that gravity could become very strong even at the relatively small energies (compared to that of the Big Bang) achievable at the LHC. In that case, the LHC collisions could produce black holes, fascinating objects of extreme matter densities, such that nothing, not even light, can escape after coming closer than a given distance.

The name "black holes" brings to the mind enormous objects of cosmic scales, like the gigantic black holes thought to be at the centre of galaxies, or swallowing the gas from an accretion disc. But because of Einstein's equation, a black hole produced in the LHC cannot have a mass larger than the collision energy divided by c^2, about 0.000000000000000000025 grams! According to a theory formulated by Steven Hawking, such small black holes would "evaporate" due to quantum fluctuations, therefore rather than swallowing the matter around them these objects would "decay" like any other heavy object produced in the LHC, and lead to collisions that only an expert could distinguish from the many others.

But what if Hawking was wrong, or if anyway other dangerous objects could be produced by colliding beams at these unprecedented energies? The answer comes from looking at the sky. LHC energies are record-breaking on Earth, but high-energy collisions of cosmic rays occur constantly in the universe. The fact that the universe has not collapsed as a consequence of these collision demonstrates the negligible danger of the LHC. After the first period of data taking, no indication of black holes production has been found. And this is scientifically a bad news, since both the possibility of producing these objects in a controlled environment, and the study of Hawking's radiation would be amazing scientific discoveries, with no dangers (especially for the physicists working at the LHC, who would be the first to be affected, and are the first to be concerned about not producing anything harmful).

Chapter 7

The Discovery of the Higgs Boson

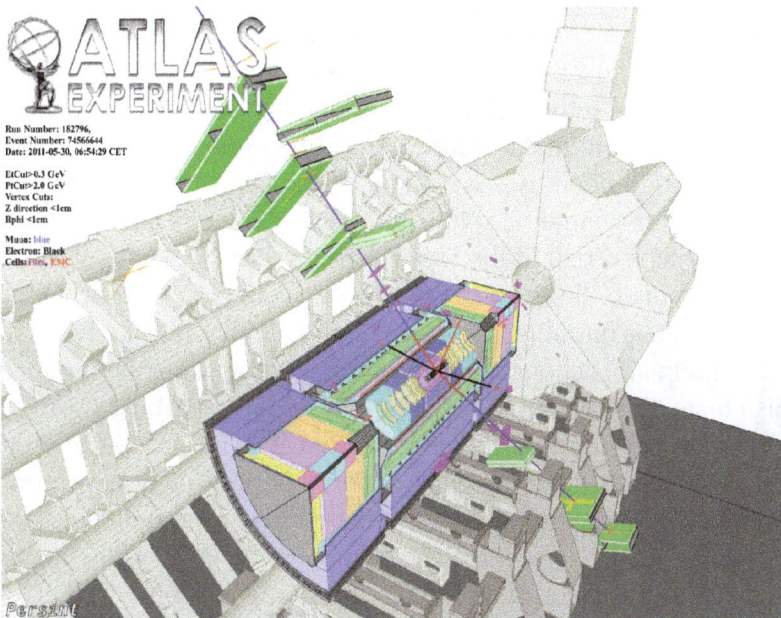

Figure 15: A Higgs boson candidate. In this decay mode, the Higgs produces two Z bosons; one of them decays into a pair of electron and antielectron, the other into a muon and an antimuon.

The Higgs boson has been for decades the Holy Grail of particle physics, the missing piece of the puzzle of the electroweak theory. This very successful model, able to explain an enormous quantity of experimental observations, is based on symmetry principles, basically stating that all fermions have the same intrinsic mass, zero. But we do know that most of the particles have nonzero mass, and actually their mass difference explains why our universe is full of electrons, but there are very few muons and taus, particles very similar to the electron but much heavier. The electroweak theory has a very elegant explanation for the mass difference between all these particles: the underlying symmetry is "broken" by an interaction with an external field, called the Higgs field.

The universe is filled by a scalar field (meaning that in each point it can be represented by a single value, like the temperature in a room; as opposed to the gravitational field that is described not only by a value but also by a direction). Particles that would otherwise be massless (and therefore always travel at the speed of light) do interact with this field, are "slowed down" by this interaction, and acquire a mass (which can be thought of as the energy of the interaction of each particle with this field; after all, mass is energy, from Einstein's equation). In more mathematical terms, the breaking of this symmetry can be described as a symmetric potential for a system that will fall into a non-symmetric solution. We can think of a perfectly round mountain with a ball at the top. When the ball is exactly at the top of the mountain, the system is perfectly symmetric under rotations. But this state is unstable: sooner or later the ball will fall off one side of the mountain, "choosing" a direction, and breaking the symmetry. This creates the masses of the elementary particles; but beware, do not blame the Higgs mechanism for your weight: protons and neutrons, which constitute the great majority of the weight of our body, are not elementary particles, and their masses are much larger than that of the quarks composing them, so in reality most of the mass of the known universe is rather due to the strong interactions.

But how can one demonstrate the validity of this model? In quantum mechanics, each field is carried by a particle (for instance, the electromagnetic field is carried by photons), and the Higgs field is no exception: it is carried by the now famous Higgs boson. Even if Higgs bosons, like the field they carry, are everywhere, producing "real" Higgs bosons is a very rare process, and requires an accelerator with very high beam intensities plus enough energy to be able to create it. After being created, the Higgs boson does immediately decay, so, like all the heavy states produced in the collider, it has to be reconstructed from its decay products. The Standard Model does not predict the mass of the Higgs, and the preferred decay modes depend on the mass, so this particle has been searched by making various assumptions on its mass, and selecting the corresponding decay modes.

Being a neutral particle, the Higgs boson can decay into a pair of a particle and its antiparticle, and since it couples to the mass, it will preferably decay into the heaviest pair energetically allowed. So for instance if the Higgs is heavier than 350 GeV, twice the mass of top quark (175 GeV), it would decay into top quark pairs; if it is between 350 and 180 GeV it would decay into Z boson pairs (the mass of the Z boson is about 90 GeV), etc. In the mass region around 120 GeV, the Higgs would preferably decay into a pair of b-quarks, a process almost impossible to observe due to the large amount of b-quarks produced by other processes. On the other hand, even though it is too light to produce a proper pair of Z bosons, due to the uncertainty principle the Higgs can, in a small number of cases, decay into a real and a "virtual" Z boson — a Z boson produced by a quantum fluctuation that immediately decays, preferably into two leptons. This decay mode is experimentally almost background-free, and is a very good channel to look for the Higgs boson production. In a similar manner, the Higgs can decay into a pair of virtual top–antitop quarks that immediately annihilate among them, producing two photons in the final state. Even though the Higgs cannot directly decay into photons because they have zero mass, the first indication of the existence of this particle came from this decay mode, and the ATLAS and CMS detectors have built very good electromagnetic calorimeters thinking about this specific Higgs search.

There are however other processes that have nothing to do with the Higgs but produce four leptons or two photons in the final state. There is no way of knowing whether a specific event with one of these topologies comes from a Higgs decay or from any of the background processes; proof for the Higgs discovery came from a complex statistical analysis of a large number of events. The most important quantity in this specific analysis is the invariant mass, which can be calculated from the energy and direction of the photons. If the two photons come from the decay of the Higgs, their invariant mass will be equal to the mass of the Higgs, otherwise it will not have any special meaning. It is expected therefore that the distribution of the invariant mass of photon pairs in data will be smooth for the background, and will show a peak in correspondence to the Higgs mass if a signal is present. This is exactly what was observed, as Figure 16 indicates: the data, represented by the black dots, have a smooth distribution and follow precisely the predicted background, apart from the region for a mass around 125 GeV; there a clear bump is visible, indicating that a new particle with that mass value has decayed into two photons. This observation, even if obtained by both experiments, is not enough to demonstrate the Higgs discovery: many

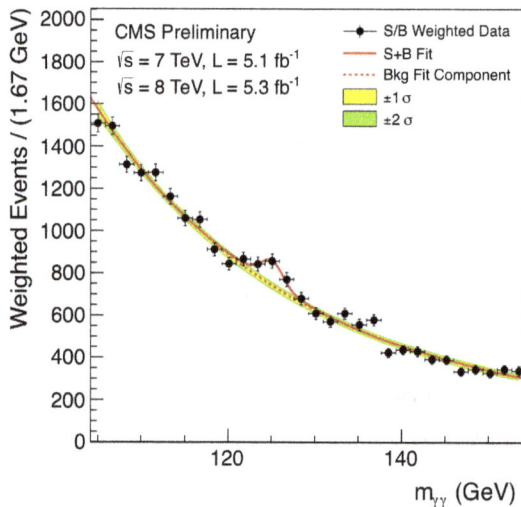

Figure 16:   Invariant mass distribution of photon pairs in CMS. Black points are the data, and the red line is the expected background. Data is compatible with the background apart in the region around 125 GeV (where the expected background is a dashed line); there a clear excess with respect to a smooth distribution is visible: a new particle with that mass can decay into two photons.

theories predict new particles that can decay into two photons. We have seen that a Higgs boson heavier than 180 GeV could decay into two Z bosons, and that thanks to the uncertainty principle, even a 125 GeV Higgs boson can, in rare cases, also decay into two Z, each of them immediately decaying into lepton–antilepton pairs. The final state will be four leptons with total charge zero, and at least one of these pairs will have an invariant mass close to that of the Z. Also in this case there are other processes that can contribute to this final state, but if the four leptons are coming from the decay of the Higgs their invariant mass will be equal to the mass of the Higgs boson, which we now know from the two-photon decay mode being equal to 125 GeV. Figure 17 shows the four-lepton invariant mass. Even though the number of events is smaller, therefore error bars are larger, a clear peak above the background is observed for a value of the invariant mass close to 125 GeV: it is the same particle that, as foreseen by the Higgs theory, decays into two photons and four leptons.

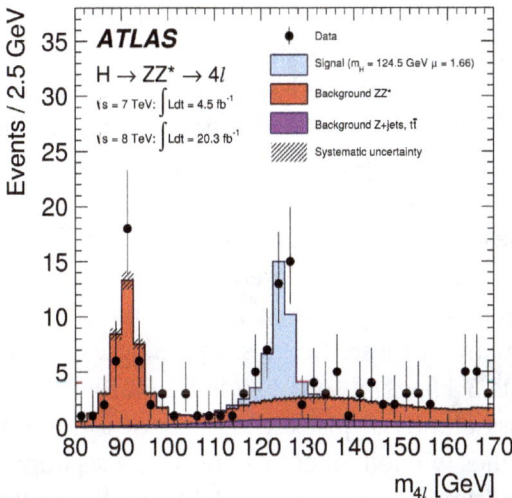

Figure 17:   The invariant mass of events with four leptons of opposite charges, as measured by ATLAS. Black points are the data, and the red histogram is the expected background. Data is compatible with the background within the vertical error bars apart in the region of invariant mass around 125 GeV, where two points are clearly higher. They are compatible with the light blue histogram, that is the theoretical prediction once a 125 GeV Higgs boson is added to the background.

Having observed the peaks in the two complementary decay modes independently between the two experiments, CERN made a public announcement on July 4th, 2012. In a super-full auditorium, with 500,000 people connected via webcast, the spokespersons of the ATLAS and CMS experiments announced to the world the discovery of a new particle compatible with the Higgs boson. And among the invited personalities were Francois Englert and Peter Higgs, two of the theoretical physicists who about fifty years before that event predicted the mechanism of mass generation. But the long wait was worth it: the two shared in 2013 the Nobel Prize in Physics. The discovery came too late for Robert Brout, co-author of Englert's article, who died a few years before being able to see his ideas confirmed by the experiment.

The discovery of the Higgs boson was just the beginning of studies of this particle for many years to come. Other measurements already followed this discovery: the decay modes in W and tau pairs have been observed; the angular distribution of the Higgs decay products confirm that this particle has spin zero (and it is the only fundamental particle with this characteristics); transverse energy and directions have been measured and compared to theoretical predictions. While all measurements performed so far leave little doubt that this particle has the properties of the Higgs, many other questions remain open. Is it really the Higgs boson predicted by the Standard Model, or does it have some different properties? Many other theories, including supersymmetry, predict one or more Higgs boson to break the electroweak symmetry, but the predicted properties are different. In many of these theories, the Higgses are more than one (SUSY predicts 5 Higgs bosons), so is the particle we found with a 125 GeV mass alone, or just the lightest of a multiplet? Time and additional studies will tell; of course the scientists hope to find deviations from the "simple" Standard Model Higgs boson that could pave the way to new discoveries.

Chapter 8

The Human Factor

Who are the scientists analysing the LHC data, studying the Higgs bosons and running the detector? Forget about the stereotype of bearded professors in white lab coats. The usual question that visitors ask when they see people taking shifts in the control room of ATLAS or CMS is: but are they scientists? They look so young, casually dressed, perhaps just hopped off their bike... Yes, they are scientists, even though perhaps a good fraction of them are PhD students, graduates who are preparing a doctoral thesis.

CERN as the host laboratory provides services to the scientific community: it employs about 2500 staff members, mainly engineers, technicians, computer scientists or administrative personnel. On the other hand, CERN only employs about 100 research physicists, about 1% of the researchers working there. The vast majority of the scientists analysing the data and running the detectors are not CERN employees, they work for universities, and usually they are not living in the Geneva area. Around one quarter of the 10,000 "users" are at CERN on an average day, but they are not always the same ones, and during special meeting periods this number will be higher.

Most of the scientists living permanently in the Geneva area are PhD students or post-doctoral researchers, who are closely interacting among themselves to carry on the analysis. Even though all meetings are performed in video conference, all results and documents posted on the web, the human interaction, being in the same room to discuss face-to-face, is still irreplaceable and it is a very formative experience for a young researcher to be at CERN and contribute to the operation of the detectors.

But how is it possible to work in a collaboration of 3000 physicists, all working for different employers, so with very few contractual obligations between them?

The collaboration has a hierarchical structure, but each scientist has freedom to work on his favourite topic. How is it possible that such a complex experiment, where also boring and tiring tasks are essential for the good performance, can run on the good will of people volunteering to do the job? A good fraction of that is due to the enthusiasm of the physicists who like their job, and their sense of responsibility, and the importance that everyone assigns to being recognised and appreciated by the colleagues. It is also true that most of the scientists do not have a permanent job for many years; out of the three thousand members of the largest collaboration, about one third are PhD students, another third is composed of post-doctoral researchers, and the last part is made of scientists with permanent academic positions.

The standard career path is to finish university around the age of 22–25 years, then start a PhD that can last between 3 and 5 years, which involves performing technical tasks as well as a full analysis and measurement. PhD students do work very hard, also because a good PhD thesis is the best passport for their future scientific career. After PhD, about half of the scientists (the other half quits research) starts a post-doctoral position, a research contract lasting from 2 to 5 years, usually in a different university (and country) from that of the PhD. These positions are very competitive, and this is clearly a factor pushing people to work hard even in the absence of formal obligations. A further hurdle comes around the age of 33–35, when scientists are considered to be mature enough to get a permanent academic position. These jobs are extremely competitive, and many scientists will at this stage quit research, others will stay all their life in a temporary contract. The small fraction of scientists obtaining a permanent position with respect to those having started a PhD explains why the average age of the researchers in an LHC experiment is so low.

Due to the mobility, the scientific community is very international, and it is not uncommon that in a university group there are more foreigners than citizens of the country where the university is located.

This makes it a vibrant and exciting environment, but it can be demanding for the personal lives of the individuals, who every few years have to move from one place to another.

Also contrary to the stereotypes, the daily life of a particle physicist is very social, since nothing can be done in isolation: both the analysis and technical tasks are performed in groups, and learning is done on the job. Also the various working group meet on a weekly basis, and since each physicist works on several different topics, various hours a day are spent in meetings where results are continuously presented and discussed. It may seem not too efficient to have all these continuous discussions, and probably in part it is not. However, in research the direction to go is never obvious, things to do and approaches to follow must be updated on a daily basis; also it is important that more senior colleagues give suggestions and ideas to the younger colleagues who are usually the ones carrying out the main load of the analysis.

Finally, there is no magic recipe. The organisation of particle physics' big collaboration is maybe not optimised in the sense of requiring a lot of work by many dedicated people, but it is very flexible and adaptable, and produced amazing results in very short time without forcing anyone to do what they did not want to.

Chapter 9

Technological Spin-Offs

The basic research made in particle physics has the aim of better understanding nature, not directly of producing technologies of practical use. The Higgs discovery for instance cannot be used to produce new forms of energy, or clean up the environment: it is "just" a deeper understanding of the concept of mass. However if we look back, often in the history of science big scientific discoveries have been followed by technological revolutions that changed the life of billions. For instance the electromagnetic unification, having understood that electricity and magnetism are indeed the same thing, is at the basis of the use of electromagnetic waves for long-distance communications, which is used for radio, TV, etc. Similarly quantum mechanics, the theory describing the behaviour of tiny particles like electrons and protons, is at the basis of all modern electronics (which is used in computers, cell phones, digital cameras, just to give a few examples), lasers, solar panels, quantum microscopes, etc.

Even if we do not know today what technological development could follow the discoveries in particle physics, we know that at CERN many technological development are under way, to prepare for the accelerators and detectors of tomorrow, and these technologies do have immediate impacts on everyday life.

The sectors where these developments are more active are electronics, computing, large scale superconductivity, precision mechanics, vacuum techniques (which can be for instance used to produce very efficient thermical solar panels), use of ultra-cold liquids, and the construction of various kinds of accelerators. The often complex requirements imposed by CERN work as a stimulus and a challenge for the industries working with the lab, and technology transfer is bi-directional.

A field that is gaining importance in the last year is medical physics. This is not a new discipline: for instance, X-rays have been used for over a century for radiography and cancer treatment. But the use of modern particle physics techniques that treat the human body as a target, has led to strong qualitative improvements in this field, which have found their way in the daily hospital practice and saved already many lives. Two technologies that are currently under development at CERN are tomographs for PET (Positron Emission Tomography) and proton therapy for cancer. PET consists of bombarding the body of a patient with small doses of antielectrons from radioactive sources; these antielectrons will annihilate with the electrons of the body, and produce photon pairs that are measured by electromagnetic calorimeters placed around the body. These calorimeters are identical to those used in LHC detectors, and the interaction of particle with the human body can be simulated using the same Monte Carlo codes used daily in particle physics.

Proton cancer therapy, or hadrotherapy, consists of bombarding cancer cells with protons, instead of using the more conventional X-rays. The advantage of protons is that they release most of their energy in a very small region of space, so it is possible to give very strong doses without risk of damaging the nearby healthy cells. This technique is very useful for tumours in very delicate areas like the brain or the eye, and for children, but requires very precise beam energy and position determination, and accurate radiation monitors, as well as simulation codes of the effects of radiation on the body, all of which is developed at CERN.

Of course the most famous technological development that came from CERN is the web, started as a collaborative tool for physicists working on the same project, and that conquered the world. So, also in the field of technological developments serendipity is around the corner, not always what is found is what was searched for. Faraday said that electric light was not invented doing research on candles; so we hope that the result of CERN's research, both fundamental and applied, could be beneficial for many more generations to come.

Conclusions

Decades of research in particle physics have led the worldwide community to embark on the most ambitious scientific project ever built, the Large Hadron Collider. Its design and construction took more than 20 years, and after four years of Run 1 the machine has restarted to take data at full collision energy.

Among the hundreds of measurements and searches performed, the discovery of the Higgs boson demonstrated that mass is the result of the interaction with an external field, clarifying the deep nature of this fundamental concept in physics. Still the study of this particle is just at the beginning, and many other open questions remain. The LHC is now running at a higher energy, focusing on the search for physics beyond the current Standard Model, with a collision energy very close to the design one.

Over the years the beam intensity will be increased by large factors, extending the discovery capabilities of the machine, but also the technological challenges of very dense particle environments. Scientists hope that finally the first hints of physics beyond what we already know will finally be observed, opening new frontiers to this field.

The quest for the fundamentals of nature will continue, as old as the human curiosity and the desire of understanding the wonderful world surrounding us.

Index

www.ingramcontent.com/pod-product-compliance
Lightning Source LLC
Chambersburg PA
CBHW070356200326

41518CB00012B/2252